Tools for Conservation



# [The Use of O]xygen-Free [Environment]s in the Control [of Museum I]nsect Pests



The Getty C[onservation Institute]

Timothy P. Whalen, *Director, Getty Conservation Institute*
Jeanne Marie Teutonico, *Associate Director, Field Projects and Conservation Science*

Chris Hudson, *Publisher*
Mark Greenberg, *Editor in Chief*
Ann Lucke, *Managing Editor*
Tevvy Ball, *Project Editor*
Sheila Berg, *Copy Editor*
Hespenheide Design, *Designer*
Pamela Heath, *Production Coordinator*

The Getty Conservation Institute works internationally to advance conservation and to enhance and encourage the preservation and understanding of the visual arts in all of their dimensions—objects, collections, architecture, and sites. The Institute serves the conservation community through scientific research; education and training; field projects; and the dissemination of the results of both its work and the work of others in the field. In all its endeavors, the Institute is committed to addressing unanswered questions and promoting the highest possible standards of conservation practice.

The Institute's Tools for Conservation series provides practical scientific procedures and methodologies for the practice of conservation. The series is specifically directed to conservation scientists, conservators, and technical experts in related fields. Previously published in the series are *Color Science in the Examination of Museum Objects: Nondestructive Procedures* (2001) by Ruth Johnston-Feller; *Infrared Spectroscopy in Conservation Science* (1999) by Michele R. Derrick, Dusan Stulik, and James M. Landry; and *Thin-Layer Chromatography for Binding Media Analysis* (1996) by Mary F. Striegel and Jo Hill.


Getty Publications
1200 Getty Center Drive
Los Angeles, California 90049-1682
www.getty.edu

Printed in the United States of America
10 9 8 7 6 5 4 3 2 1

**Library of Congress Cataloging-in-Publication Data**
Maekawa, Shin, 1952-
Oxygen-free environments in the control of museum insect pests / Shin Maekawa, Kerstin Elert.
p. cm.
Includes bibliographical references and index.
ISBN 0-89236-693-1 (pbk.)
1. Buildings--Pest control. 2. Museums--Protection. 3. Insect pests.
4. Gases, Asphyxiating and poisonous. 5. Museum conservation methods.
I. Elert, Kerstin. II. Title.
TH9041 .M34 2003
069'.53--dc21 2002154104

# Contents

# Foreword

For more than fifteen years, scientists at the Getty Conservation Institute have been actively conducting research on solutions to the problem of pest control in museum environments and have collaborated with partners all over the world in the effort to house documents and objects in safe, protective environments. While this work was originally undertaken in an effort to effectively house the royal mummies at the Egyptian Museum in Cairo, other potential uses of the technology became apparent, and the ultimate result of that research and work is this present volume.

This publication, in the Tools for Conservation series, builds on the research published by the GCI in 1998 in volumes in the Research in Conservation series: *Inert Gases in the Control of Museum Insect Pests* and *Oxygen-Free Museum Cases*. This book turns the theory of those two publications into practice, providing substantive practical and specific information about the use of oxygen-free environments (in this case, nitrogen) in integrated pest management in the museum environment. The first four chapters provide background materials, and the fifth chapter provides step-by-step instructions on procedures as well as on the construction of treatment systems for insect anoxia.

A searchable CD-ROM in the back flap of the book is intended to help make this publication particularly easy and practical to use. URLs and web information are provided. Comments from readers regarding the usefulness of the CD-ROM are welcome, as the goal of the book is to create a useful tool for the practitioner in the field.

The book was written by Shin Maekawa, senior scientist at the Getty Conservation Institute, and Kerstin Elert, of the University of Granada in Spain. The work benefited from technical editing by Frank Lambert and from developmental editing by Elizabeth Maggio, who was especially helpful in shaping Chapter 5, on protocols. Sheila Berg was the copy editor; the project editor was Tevvy Ball. All the editorial staff involved in the project have greatly enhanced the quality of the final publication. The volume was designed by Hespenheide Design, and production coordination was provided by Pamela Heath.

*Timothy P. Whalen*
*Director*
*The Getty Conservation Institute*

# Preface

This book describes the methods that we recommend for eradicating museum insect pests. A previous volume from the Getty Conservation Institute, *Inert Gases in the Control of Museum Insect Pests*, presented a broad spectrum of approaches, procedures, and conservators' experiences. However, individuals with limited resources and many responsibilities may want more specific information to aid them more directly in solving their insect infestation problems.

In this book we use nitrogen as the inert gas to replace oxygen and a limited number of types of containers as treatment chambers. We describe in detail the procedures for fabrication of enclosures and for their use in anoxia. This book is practical rather than theoretical and specific rather than general. It is written to aid conservators and others to successfully eradicate insects in valued collections by means of placing infested objects in oxygen-free environments for a specified period. (In this book "oxygen-free environment" is meant as a practical working term, referring to an atmosphere that contains less than 0.3% oxygen rather than one that literally has 0% oxygen content.)

The detailed step-by-step instructions in Chapter 5 may be unnecessary for individuals who are skilled in anoxia treatment. It is our hope, however, that this book will allow these experts to hand over the task of insect anoxic treatment to conscientious novices, with minimal guidance needed to assure success.


## Acknowledgments

We thank Frank Lambert, Professor Emeritus at Occidental College, for his valuable advice throughout the ten years of research on oxygen-free environments for conservation at the Getty Conservation Institute and for serving as technical editor of this publication.

# Introduction

Insects, especially moths and beetles, are major threats to museum objects made of organic materials. Over the years they have grievously damaged or destroyed countless irreplaceable artifacts. Museum personnel worldwide struggle with the complex problem of infestation, made more complex by the need to avoid risks to artifacts and to personnel posed by hazardous pesticides. In recent years, after the harmful effects on humans of pesticides such as DDT and Lindane and fumigants such as ethylene oxide were found, safer control methods in museums have become increasingly necessary.

Today it is agreed that successful pest control in museums requires an integrated approach that includes assessment, preventive procedures, and active treatment. Several manuals and handbooks are available that provide detailed information on the individual components of an integrated pest management program (Harmon 1993; Pinniger 2001; Pinniger and Winsor 1998; Story 1985; Zycherman and Schrock 1988). Active treatments are commonly classified according to the type of application and the chemicals used and include fumigants, dusts, and sprays.

Obviously, according to the principle "Do no harm to an object," the application of dusts and sprays is suspect because they leave residual material. Although this residue may be beneficial in continuing to kill insects, it or its long-term degradation products have the potential to damage objects over time. Fumigants have appeared less threatening to treating museum collections in that they do not leave residues. However, relatively recent information indicates that widely used fumigants such as methyl bromide and sulfuryl fluoride (Vikane) may be deleterious to paintings and art materials (Koestler et al. 1993). Ethylene oxide, a fumigant once adopted by major museums and other institutions, poses an especially serious personnel hazard, in addition to reacting with chemical groups on cellulosic art objects.

Low-oxygen atmospheres have been employed successfully to control pests in food storage and have been increasingly adopted as a safer alternative to conventional fumigation with hazardous chemicals. The method is simple and can be modified according to the specific needs of the museum, for example, the size and number of objects, their varied humidity requirements, and the frequency of recurrent infestation. However, currently available publications lack specific recommended procedures for the wide variety of applications and problems encountered in anoxic disinfestation.

This book is designed to support those who are beginning anoxic treatment of insects. The first four chapters provide background material, as follows. Chapter 1 briefly surveys pest control but primarily emphasizes the factors that determine the mortality of insects in inert gases. Chapter 2 discusses the two basic techniques for attaining nitrogen atmospheres: (1) removal of oxygen from the air in treatment containers by oxygen absorbers and (2) flushing the air from the container with nitrogen. It also discusses the choice and use of oxygen-barrier films as container materials and the various adjunct devices needed for anoxic procedures. Chapter 3 deals with procedures for small-scale anoxic treatment and the types and behavior of oxygen absorbers. Most important are the specific equations for calculating the amount of oxygen absorber needed as well as the amount of silica gel that is essential for buffering the relative humidity in bags. Chapter 4 is concerned with large-scale anoxic treatment. It begins with the three types of nitrogen supply, then develops details of the in-house fabrication of large enclosures, the adaptation of large fumigation tents, or "Bubbles," for anoxia, and the use of an automated rigid chamber.

Chapter 5 is a detailed, how-to guide for fabricating and using containers for anoxic treatments. These containers range from small pouches to large commercial fumigation tents to rigid chambers such as those previously employed by some museums for fumigating collections. The list of six basic types of museum situations, involving both large and small objects, permits readers to choose a practical protocol for solving a specific insect problem.

## References

**Harmon, J.**
1993. *Integrated Pest Management in Museum, Library and Archival Facilities.* Indianapolis: Harmon Preservation Pest Management.

**Koestler, R. J. E., E. Parreira, E. Santoro, and P. Noble**
1993. Visual effects of selected biocides on easel painting materials. *Studies in Conservation* 38:265–73.

**Pinniger, D.**
2001. *Pest Management in Museums, Archives, and Historic Houses.* London: Archetype Publications.

**Pinniger, D., and P. Winsor**
1998. *Integrated Pest Management.* London: Museums and Galleries Commission.

**Story, K. O.**
1985. *Approaches to Pest Management in Museums.* Suitland, Md.: Smithsonian Institution Press.

**Zycherman, L. A., and J. R. Schrock**
1988. *A Guide to Museum Pest Control.* Washington, D.C.: Foundation of the American Institute for Conservation of Historic and Artistic Works and Association of Systematics Collections.

Chapter 1

# Insect Mortality Using Anoxia

In this chapter we summarize current knowledge of the lethal action of oxygen-free atmospheres on insects and discuss its implications for treating museum artifacts. We then discuss the factors that most influence the effectiveness of anoxia treatments: oxygen content, relative humidity, and temperature. Finally, we describe the research of Rust and Kennedy (1993, 1995), which forms the basis for anoxic treatment recommended in this book.

## Pest Control in Museums

Over the past decade many museums have increased their use of inert gases to eliminate insects from infested objects. (See Table 1.1 for typical insect pests in museums, archives, and cultural institutions.) The reasons are clear. Practical experience has shown that conventional fumigants can have adverse effects on various materials, and some fumigants or their reaction products are retained in the artifacts for long periods (Florian 1987). Further, they pose threats to human health and must be applied by trained personnel or contracted to specialized pest control operators. Some fumigants, such as methyl bromide, have been judged environmentally harmful, and their use is restricted by governmental agencies in many countries.

An alternative to fumigating artifacts is to expose them to very low or high temperatures. However, research on the effect of exposure to extreme temperatures has not dispelled concerns regarding the safety of all objects under such conditions. Thus this treatment is limited to museum objects that contain thermally robust materials.

Atmospheres with low concentrations of oxygen had been used for the control of pests in stored food products for several decades before this method was adapted to the treatment of museum artifacts in the late 1980s. Considerable research on the efficiency of modified atmospheres for the control of pests in stored food has been conducted since about 1970, and findings have been published in numerous articles. Useful reviews of this early work are given by Bailey and Banks (1975, 1980).

To translate the success of industrial efforts to the field of museum pests, researchers proposed using several gases, including nitrogen, argon, and carbon dioxide. However, carbon dioxide in

**Table 1.1**

Typical insect pests in museums

| Common Name | Scientific Name |
|---|---|
| American cockroach | *Periplaneta americana* |
| Australian spider beetle | *Ptinus ocellus* |
| Black carpet beetle | *Attagenus piceus* |
| Book beetle | *Nicobeum casteneum* |
| Book louse | *Liposcelis corrodens* |
| Bread beetle | *Stegobium paniceum* |
| Brownbanded cockroach | *Supella longipalpa* |
| Cabinet beetle | *Trogoderma inclusum* |
| Casemaking clothes moth | *Tinea pellionella* |
| Cigarette beetle | *Lasioderma serricorne* |
| Common clothes moth | *Tineola bisselliella* |
| Confused flour beetle | *Tribolium confusum* |
| Deathwatch beetle | *Xestobium rufovillosum* |
| Drugstore beetle | *Stegobium paniceum* |
| Drywood termite | *Cryptotermes brevis* |
| Firebrat | *Thermobia domestica* |
| Fruit fly | *Drisopholus melanogaster* |
| Fur beetle | *Attagenus fasciatus* |
| Furniture beetle | *Anobium punctatum* |
| Furniture carpet beetle | *Anthrenus flavipes* |
| German cockroach | *Blattella germanica* |
| Grain mite | *Acarus siro* |
| Hide beetle | *Dermestes maculates* |
| Khapra beetle | *Trogoderma granarium* |
| Larder beetle | *Dermestes lardarius* |
| Longhorn borer beetle | *Hylotrupes bajulus* |
| Merchant grain beetle | *Oryzaephilus mercator* |
| Mold mite | *Tyrophagus putrescentia* |
| Odd beetle | *Thylodrius confusum* |
| Old house borer | *Hylotrupes bajulus* |
| Powderpost beetle | *Lyctus* spp. |
| Red flour beetle | *Tribolium castaneum* |
| Redlegged ham beetle | *Necrobia rufipes* |
| Rice weevil | *Sitophilus oryzae* |
| Silverfish | *Lepisma saccharina* |
| Varied carpet beetle | *Anthrenus verbasci* |
| Webbing clothes moth | *Tineola bisselliella* |

concentrations in excess of 1.5% are reported to produce a loss of consciousness and lead to death (Air Products 2001).[1] The maximum permitted short-term exposure level established by the Occupational Safety and Health Administration (OSHA) is 0.5% (5000 ppm). This level easily might be reached if leakage occurs during large-scale treatments, given that carbon dioxide concentrations of about 60% are required for the control of insects. Valentin (1993) has examined the use of argon rather than nitrogen to eliminate museum insect pests. Her data indicate that argon reduced the minimum exposure time for insect mortality in many cases, but in at least one case, involving *Hylotrupes bajulus* (L.), at 20°C there was no time reduction. Because a limited number of insects have been tested under limited conditions with argon, decisions about exposure time involve significant guesswork. Thus argon should perhaps be considered a guarantor of insect elimination rather than a time saver over nitrogen. In pest management of valuable or priceless objects the goal always should be complete elimination of insects rather than saving time. Only nitrogen will be used in the procedures discussed in this book because of the extensive and conclusive results of Rust and Kennedy (1993, 1995).

## Major Factors Influencing Insect Mortality

Although the efficacy of low-oxygen atmospheres in eliminating insects is not questioned, the exact physiological mechanism of nitrogen anoxia is not completely understood. Entomologists studied the behavior of insects in modified atmospheres and found that the sensitivity to low oxygen concentrations varied with species, developmental stage, and the environmental conditions to which the insects were exposed before as well as during the anoxia treatment. Generally, mobile life-forms such as adults and larvae were more susceptible than the relatively inactive forms, pupae and eggs. AliNiazee (1972) reported that the active life stages of *Tribolium confusum* (Jacquelin du Val) and *Tribolium castaneum* (Herbst)[2] became hyperactive when first exposed to low-oxygen atmospheres but were then anesthetized, whereas the inactive stages showed no visible change. Most insects are capable of restoring all body function after an exposure to low-oxygen atmospheres for several hours, sometimes even for days, but die if the exposure time is extended.

Researchers agree that desiccation is an important mechanism influencing insect mortality during exposure to low-oxygen atmospheres.

Jay and Cuff (1981) demonstrated that the weight loss of different stages of *Tribolium castaneum* (H.) was 1.6 to 4.8 times higher after exposure to low-oxygen atmospheres (99% nitrogen) than after exposure to air under identical humidity and temperature conditions. They therefore suggested that water loss was the major cause of death.

The process of water loss is closely linked to respiration, the exchange of oxygen and carbon dioxide. An insect's respiratory (tracheal) system has openings to the atmosphere (spiracles) that can be partially or completely closed to prevent water loss (Blum 1985). This feature allows insects to maintain a sufficient water reserve in environments with low relative humidity. However, air containing low oxygen levels, about 2%, or sufficient concentrations of carbon dioxide above 10% cause the spiracles of a well-hydrated insect to open fully, resulting in a higher rate of water expiration (Bursell 1970). After a critically low moisture content is reached, the insect tends to reduce its oxygen consumption to minimize further water loss. Of course, reduced oxygen consumption could enable the insect to survive anoxic conditions for an extended period (Gilberg 1989).

It has been proposed that insects rely on anaerobic metabolism as a last resort to survive anoxic conditions. However, death will still occur in time because of the insect's consequent accumulation of abnormal quantities of various metabolic end products (AliNiazee 1972). More recent studies by Adler (1994) confirm that the accumulation of lactate in insects during exposure to a nitrogen atmosphere leads to acid imbalance in an insect's body. This seriously inhibits the anaerobic metabolic pathway.

### Temperature

Insects generally show increased metabolic rates and consume more oxygen when the temperature is raised. When a certain critical temperature is reached, the insect might use evaporative cooling by opening its spiracles to control body temperature, thus promoting increased water loss (Bursell 1970). These two phenomena may explain the greatly increased mortality rate of insects when anoxia treatments are carried out at higher temperatures. Several studies showed that raising the temperature from 15°C to 35°C can reduce the required exposure time for death by 25% to 65% for each 5°C increase (AliNiazee 1972; Lindgren and Vincent 1970; Navarro and Calderon 1980; Soderstrom, Mackey, and Brandl 1986; Rust and Kennedy 1995) (Fig. 1.1).

Temperatures above 35°C seem to cause a drastic increase in mortality rate beyond what would be expected to be caused by the low-oxygen atmosphere alone (Marzke, Press, and Pearman 1970). However, it must be emphasized that these mortality rates vary with species, sometimes even with the developmental stage, and they are also influenced by temperature conditions to which the insect has previously been exposed. Acclimation to higher temperatures can increase the critical temperature insects are capable of surviving (Bursell 1970).

A combination of low-oxygen atmospheres and elevated temperatures of between 25° and 55°C, depending on the artifacts' sensitivity

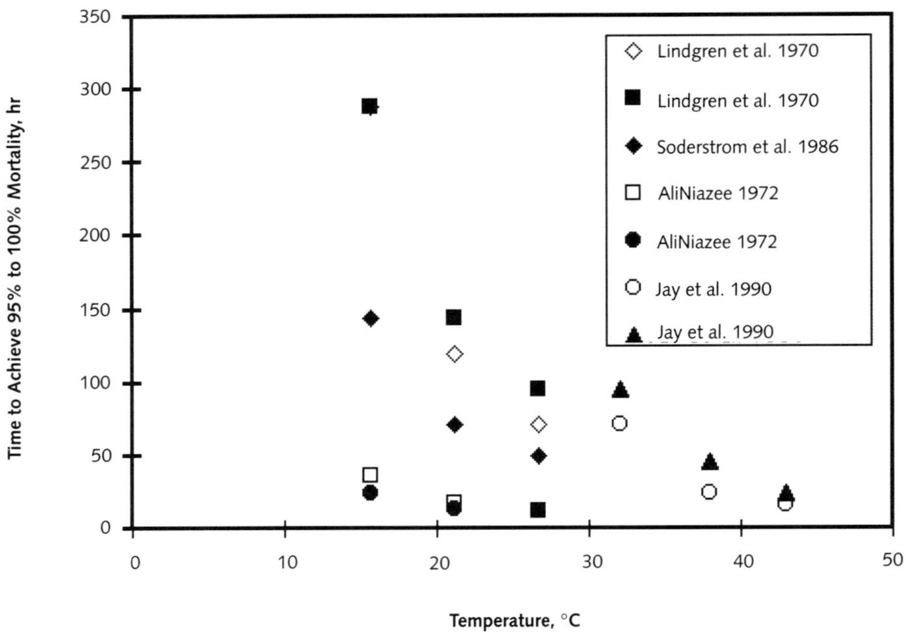


**Figure 1.1**
Influence of temperature on insect mortality in nitrogen atmospheres at oxygen concentrations of 1% or below

to heat, has been proposed for the treatment of insect-infested objects (Thermo Lignum Ltd. 1995). Extensive earlier work by Valentin (1993) had thoroughly established this point. Of course, conservators are keenly aware of the potential harm to some items in their collections if they were to be exposed to the low relative humidity environments caused by high temperatures. (Humidity is controlled in the Thermo Lignum process.)

### Relative humidity

Insects lose water during exposure to low-oxygen atmospheres via evaporation as described earlier. The rate of evaporation depends on the ambient humidity level and will increase in a dry environment (Bursell 1970). Soderstrom, Mackey, and Brandl (1986) demonstrated that, depending on the species, complete mortality can be achieved in 10% to 30% less time if the relative humidity is decreased from 60% to 40% during exposure to a low-oxygen atmosphere. Valentin and Preusser (1990) reported a decrease of 40% to 50% in the time required to eliminate the insects, depending on developmental stage, when the relative humidity was reduced from 75% to 40% (Fig. 1.2). Although Rust and Kennedy (1995) could not observe a noticeable decrease in the exposure time when the relative humidity was reduced from 75% to 55%, they confirmed Valentin and Preusser's reduction of exposure time when it was lowered from 55% to 33%.

### Oxygen concentration

Studies of the influence of oxygen concentration on the mortality of pests in stored food products show that mortality rates increase when oxygen concentrations are lowered. Atmospheres containing more than 2% oxygen are not effective in killing insects of many species (AliNiazee 1971; Marzke, Press, and Pearman 1970; Soderstrom, Brandl, and Mackey, 1992; Soderstrom, Mackey, and Brandl 1986). Rust and Kennedy (1995) investigated the influence of oxygen concentration on the mortality of two common museum insects, *Lasioderma*

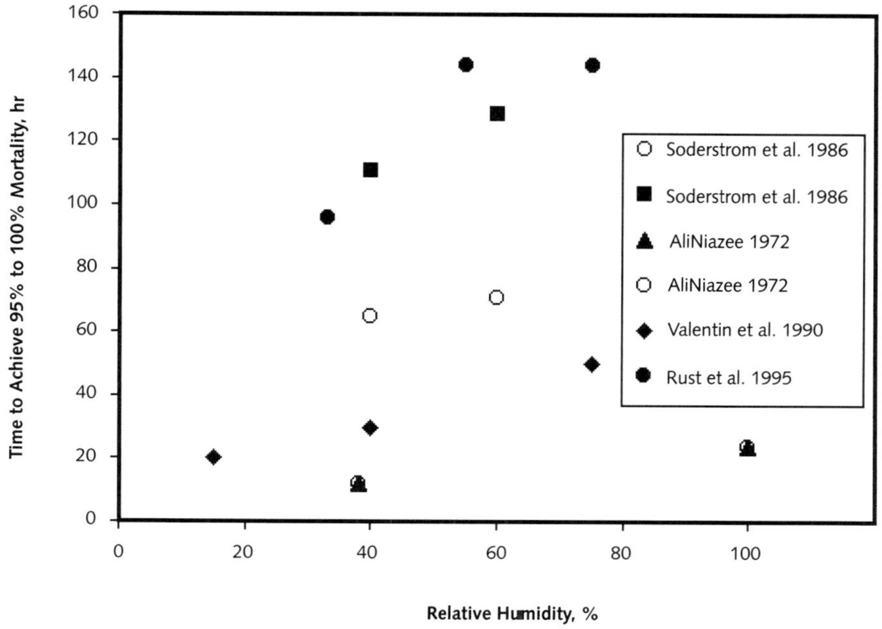


**Figure 1.2**

Influence of relative humidity on insect mortality in nitrogen atmospheres at oxygen concentrations of 1% or below

*serricorne* (Fabricus) and *Anthrenus flavipes* (LeConte), in a study sponsored by the Getty Conservation Institute (GCI). They tested the effectiveness of nitrogen atmospheres containing oxygen concentrations of 0.3, 0.6, and 0.95% at a temperature of 25.6°C and found that complete mortality of all stages of both species was achieved after seven days in an atmosphere containing 0.3% oxygen at relative humidities between 33 and 75%. However, the necessary exposure time had to be increased by 50% to 200% to achieve complete mortality in 0.6% oxygen concentration. Further, even using a greatly extended exposure time, they could not obtain complete mortality at 0.95% oxygen. This indicates that anoxia treatments would need to be lengthened to more than three weeks to ensure complete eradication of all life stages of insects at oxygen concentrations higher than 0.3% (Fig. 1.3).

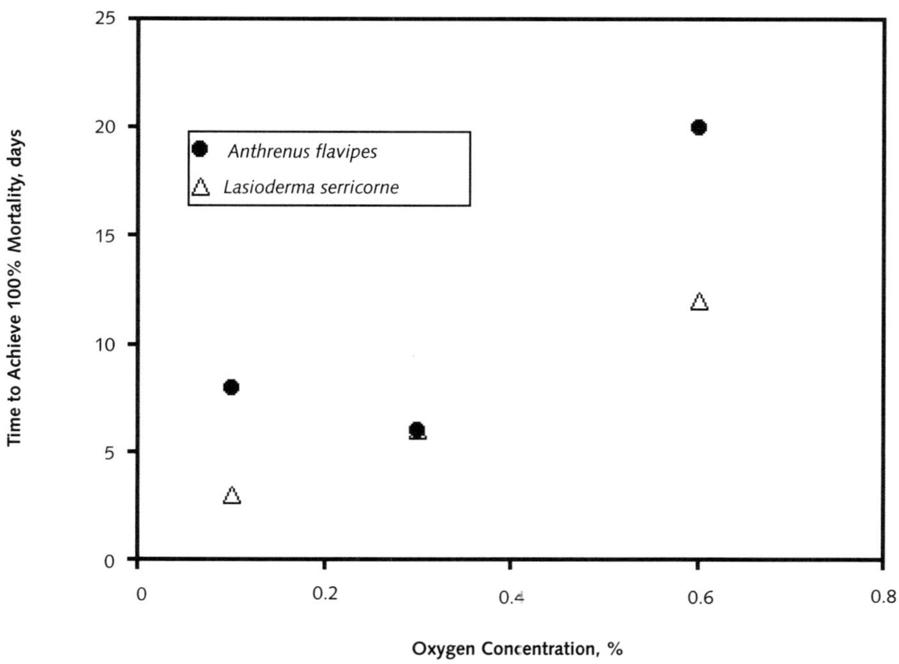


**Figure 1.3**

Exposure time required to achieve 100% mortality of all life-forms in atmospheres with varying oxygen concentrations (Rust and Kennedy 1995)

It should be emphasized that the easily visible adult stages of insects are most quickly eliminated in a low-oxygen environment. Thus the presence of dead adults during or soon after an anoxia treatment should not be considered evidence of thorough eradication. The often invisible eggs are the most resistant to anoxia and can appear as adults in a too briefly treated object many weeks later.

### Combination of low-oxygen atmospheres and carbon dioxide

Bailey and Banks (1980) reported that the presence of a few percent of carbon dioxide accelerates the lethal action of low-oxygen atmospheres. This might be explained by a higher rate of water loss (Bursell 1970) or by increased acidification in the insects' bodies (Adler 1994). Carbon dioxide concentrations of 10% to 15% have been shown to produce substantial mortality in atmospheres where the oxygen concentration was too high to be lethally effective in the given exposure time (Calderon and Navarro 1980). Rust and Kennedy (1995) showed that the required exposure time to achieve 100% mortality of *Lasioderma serricorne* (F.) eggs decreased by 40% when 5% carbon dioxide was added to a nitrogen atmosphere containing 0.62% oxygen at 55% RH and 25.6°C. However, this is a difference of only ten days versus six days. Most significant for a real-world situation, there was no difference in mortality rates of any of the life stages of *Lasioderma serricorne* (F.) under the preceding conditions when the beetles were deeply buried in four inches or more of flour.

## Optimal Parameters for the Anoxia of Museum Pests

In the available studies on the mortality of insects in low-oxygen atmospheres, there are inconsistencies in the time needed to achieve 100% mortality. Two main factors cause these differences. First, variations in the test protocol among researchers—for example, differences in temperature or relative humidity, lack of accurate control of oxygen concentration over the entire test period—result in major changes in the lethality time. Second, variations among the insects themselves—for example, age, rearing conditions, stress resulting from handling, availability of nutrition—can change their sensitivity to low-oxygen atmospheres. In some studies only two or three susceptible species or a limited number of individuals of several species may have been used, resulting in a misleading outcome.

The majority of mortality studies of pests in stored food or grain have been done at 1% oxygen concentration or above because lower concentrations are generally difficult to maintain in huge commercial storage facilities. Further, the stored food industry can accept 99% mortality for economic reasons. However, only mortality data obtained from tests carried out below 1% oxygen concentration are considered in this book. Anoxia treatments performed at higher oxygen concentrations may not produce 100% kill or may require higher temperatures to achieve total mortality of insects and therefore are not generally practical or suitable for museum objects.

To recapitulate some results from previous studies so that they can be compared: Gilberg (1989) reported 100% mortality of *Tineola bisselliella* (Hummel), *Stegobium paniceum* (Linnaeus), and *Anthrenus vorax* (L.) when exposed to 0.4% oxygen concentration at 60% to 70% RH and 30°C for seven days. Rust and Kennedy (1993) evaluated the mortality of all stages of twelve common museum pests exposed to 0.1% oxygen in nitrogen at 55% RH and 25.5°C and found that 100% mortality was achieved for most insects after five days.[3] However, larvae and eggs of the *Lasioderma serricorne* (F.) required six and eight days respectively (Fig. 1.4). In an extension of their study Rust and Kennedy (1995) tested the effectiveness of nitrogen atmospheres containing higher oxygen concentrations of 0.3, 0.6, and 0.95% at a temperature of 25.6°C. Complete mortality of all stages of *Lasioderma serricorne* (F.) and *Anthrenus flavipes* (L.) was achieved after seven days in an atmosphere containing 0.3% oxygen at relative humidities of between 33 and 75%. Exposure periods of more than twenty days were necessary to achieve complete mortality for all stages of both species at 0.6% oxygen concentration. Valentin (1993) exposed all life stages of eight common museum pests[4] to a low-oxygen environment in nitrogen containing 0.03% oxygen at 40% RH and temperatures of 20°, 30°, and 40°C. Complete mortality was achieved for most insects after nine days at 20°C and six days at 30°C. Only *Hylotrupes bajulus* (L.) required twenty days' exposure at 20°C and ten days at 30°C. At a temperature of 40°C it was possible to eradicate all species within one day.

One well might be concerned that insects hidden deep inside an object could survive over longer periods than under common test conditions, in which insects are kept in open containers from which the oxygen can rapidly be removed. Rust and Kennedy (1993) were able to demonstrate that *Incisitermes minor* (Hagen) were eradicated as quickly when enclosed in wood blocks as they were when kept in containers with rearing media; in both cases it took three days to achieve complete mortality of *Incisitermes minor* (H.) exposed to 0.1% oxygen concentration at 25.5°C and 55% RH. Studies at the Getty Conservation Institute

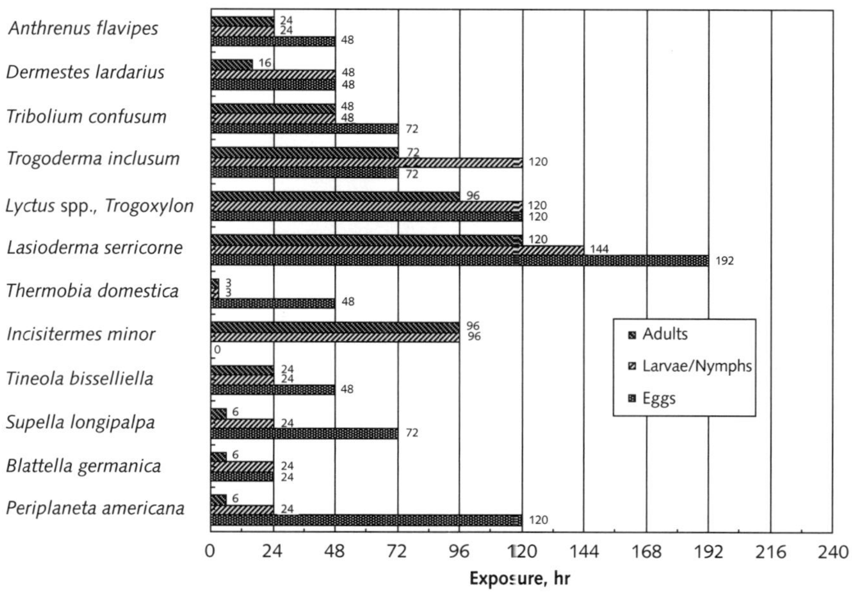


**Figure 1.4**
Exposure time required to achieve 100% mortality of each life-form of twelve common museum pests in an atmosphere with 0.1% oxygen at 25.5°C and 55% RH (Rust et al. 1996)

conducted by Maekawa and Lambert showed that it took about two days for the oxygen in coated and uncoated solid wood blocks (poplar and oak, 3.8 × 3.8 × 70.4 cm) to be displaced with nitrogen, but less than four hours for it to be displaced in termite-infested wood. This indicates that after two days anoxic conditions will be established inside solid wood and much sooner in infested wood.

From the above test results, it can be concluded that all life-forms of the majority of insect species can be eradicated within eight days at oxygen concentrations of less than 0.3% at 50% RH and 25°C. A person responsible for the control of pests in museum objects must obtain the total elimination of any insects in an anoxia treatment. Thus a period of fourteen days under the above conditions is advisable for assurance of complete kill. This would include about two days for establishing anoxic conditions in the enclosure, plus a safety margin of some 50% more than the experimentally proved survival time of common pests. However, at temperatures below 25°C there must be an increase in treatment time of about 50% for each 5°C decrease; that is, about twenty-two days will be required at 20°C. The most resistant insects may not be killed in twenty-two days at 20° C. (See Table 1.2 for recommended treatment times at 20°C and 25°C for specific species.) It is risky to perform anoxic treatments at temperatures below 20°C (see Fig. 1.1). In cold conditions the installation of space heaters is recommended to maintain the temperature at 20°C, preferably at 25°C. Alternatively, scheduling anoxic treatments for the warmer season of the year should be considered.

Valentin's (1993) research proves that the time needed to kill invading insects can be significantly reduced by raising the temperature of the anoxic treatment to 40°C. However, it might be difficult to maintain such elevated temperatures when large-scale treatments are per-

**Table 1.2**

Recommended time for anoxia treatment at 20° and 25°C at oxygen concentrations less than or equal to 0.3%

| | Recommended Treatment Time (days) | |
|---|---|---|
| **Pest Species** | ***at 20°C*** | ***at 25°C*** |
| *Tineola bisselliella* | 7 | 5 |
| *Anthrenus flavipes* | 7 | 5 |
| *Thermobia domestica* | 7 | 5 |
| *Trogoderma inclusum* | 14 | 10 |
| *Dermestes lardarius* | 7 | 5 |
| *Lasioderma serricorne* | 22 | 14 |
| *Tribolium confusum* | 10 | 7 |
| *Periplaneta americana* | 14 | 10 |
| *Supella longipalpa* | 10 | 7 |
| *Blattella germanica* | 5 | 4 |
| *Lyctus* spp. and *Trogoxylon prostomoides* | 14 | 10 |
| *Incisitermes minor* | 11 | 8 |

*Note:* The times given for anoxic elimination of the species on each line are solely for infestation by that particular species. If the identity of insects is not known or if several types are present, the most prudent course is to follow the general recommendation of 14 days of anoxia at 0.3% or less of oxygen at 55% RH and 25°C. At the same conditions but 20°C, the time should be 22 days.

formed. Even more important, heat-sensitive and RH-sensitive materials in museum objects might be damaged.

If the exposure time is reduced below the recommended fourteen-day minimum, it is important to identify the most tolerant species present in the infested material. Some insects such as *Hylotrupes bajulus* exhibit exceptional resistance to low oxygen concentrations and require unusually long exposures. In contrast, moths such as *Tineola bisselliella* (Hummel) are usually eradicated within a few days. Assistance in the identification of insect species may be obtained from the entomology departments of universities or from large pest control companies. When applying available information from insect mortality studies to museum practice, it is essential to match the research or published conditions with the museum treatment conditions so that none of the parameters of temperature, relative humidity, or oxygen concentration differ significantly.

## Notes

1. The hazards as well as the safe handling of carbon dioxide—and of gaseous and liquid nitrogen that are more important in most insect anoxia—are well presented in the Safetygrams of Air Products and Chemicals, Inc. They are available directly in easily readable form at http://www.apci.com/productstewardship/product_safety/safetygrams/index.asp. Carbon dioxide is discussed in Safetygram-18, gaseous nitrogen in Safetygram-2, and liquid nitrogen in Safetygram-7.
2. See Table 1.1 for the common and scientific names of common insect pests.
3. The twelve insect species were *Tineola bisselliella* (Hummel), *Anthrenus flavipes* (LeConte), *Thermobia domestica* (Packard), *Trogoderma inclusum* (LeConte), *Dermestes lardarius* (L.), *Lasioderma serricorne* (F.), *Tribolium confusum* (Jacquelin du Val), *Periplaneta americana* (L.), *Supella longipalpa* (F.), *Blattella germanica* (L.), *Lyctus* ssp., *Trogoxylon prostomoides* (Gorham), and *Incisitermes minor* (Hagen).
4. *Hylotrupes bajulus* (L.), *Anobium punctatum* (DeGeer), *Stegobium paniceum* (L.), *Xestobium rufovillosum* (DeG.), *Attagenus piceus* (Oliver), *Nicobium castaneum* (Oliv.), *Lyctus brunneus* (Stephen), *Lasioderma serricorne* (F.)

## References

**Adler, C. S.**

1994. Carbon dioxide—more rapidly impairing the glycolytic energy production than nitrogen? *In Proceedings of the 6th International Working Conference on Stored-Product Protection*, ed. E. Highley, 7–10. Canberra: C.A.B. International.

**Air Products and Chemicals, Inc.**

2001. *Safetygram-18 Carbon Dioxide.* http://www.apci.com/productstewardship/product_safety/safetygrams/index.asp

**AliNiazee, M. T.**

1971. The effect of carbon dioxide gas alone or in combinations on the mortality of *Tribolium castaneum* (Herbst) and *T. confusum* (du Val) (Coleoptera, Tenebrionidae). *Journal of Stored Products Research* 7:243–52.

1972. Susceptibility of the confused and red flour beetles to anoxia produced by helium and nitrogen at various temperatures. *Journal of Economic Entomology* 65(1):60–64.

**Bailey, S. W., and H. J. Banks**

1975. The use of controlled atmospheres for the storage of grain. In *1st International Working Conference on Stored-Product Entomology*, 362–74. Savannah, Ga.: N.p.

1980. Review of recent studies of the effects of controlled atmospheres on stored product pests. In *International Symposium on Controlled Atmosphere Storage of Grains*, ed. J. Shejbal, 101–18. Amsterdam: Elsevier.

**Blum, M. S.**

1985. *Fundamentals of Insect Physiology*. New York: Wiley.

**Bursell, E.**

1970. *An Introduction to Insect Physiology*. New York: Academic Press.

**Calderon, M., and S. Navarro**

1980. Synergistic effect of carbon dioxide and oxygen mixtures on two stored grain insect pests. In *International Symposium on Controlled Atmosphere Storage of Grains*, ed. J. Shejbal, 79–84. Amsterdam: Elsevier.

**Florian, M. L.**

1987. The effect on artifact materials of the fumigant ethylene oxide and freezing used in insect control. In *ICOM Committee for Conservation, 8th Triennial Meeting*, Sydney, 199–208. Los Angeles: Getty Conservation Institute.

**Gilberg, M.**

1989. Inert atmosphere fumigation of museum objects. *Studies in Conservation* 34:80–84.

**Jay, E. G., H. J. Banks, and D. W. Keever**

1990. Recent developments in controlled atmosphere technology. In *Fumigation and Controlled Atmosphere Storage of Grain: Proceedings of an International Conference*, ed. R. B. Champ, E. Highley, and H. J. Banks, 134–43. ACIAR Proceedings No. 25. Singapore, February 14–18, 1989.

**Jay, E. G., and W. Cuff**

1981. Weight loss and mortality of three life stages of *Tribolium castaneum* (Herbst) when exposed to four modified atmospheres. *Journal of Stored Products Research* 17:117–24.

**Lindgren, L. D., and L. E. Vincent**

1970. Effect of atmospheric gases alone or in combination on the mortality of granary and rice weevils. *Journal of Economic Entomology* 63(6):1926–29.

**Marzke, F. O., A. F. Press, and G. C. Pearman**

1970. Mortality of the rice weevil, the Indian-meal moth, and *Trogoderma glabrum* exposed to mixtures of atmospheric gases at various temperatures. *Journal of Economic Entomology* 63(2):570–74.

**Navarro, S., and M. Calderon**

1980. Integrated approach to the use of controlled atmospheres for insect control in grain storage. In *International Symposium on Controlled Atmosphere Storage of Grains*, ed. J. Shejbal, 73–78. Amsterdam: Elsevier.

**Reichmuth, C.**

1986. Low oxygen control to control stored product insects. In *Proceedings of the 4th International Working Conference on Stored-Product Protection*, ed. E. Donahaye and S. Navarro, 194–207. Tel Aviv: N.p.

**Rust, M. K., V. Daniel, J. R. Druzik, and F. D. Preusser**

1996. The feasibility of using modified atmospheres to control insect pests in museums. *Restaurator* (Denmark) 17:43–60.

**Rust, M. K., and J. M. Kennedy**

1993. *The Feasibility of Using Modified Atmospheres to Control Insect Pests in Museums.* Getty Conservation Institute Scientific Program Report. Los Angeles: Getty Conservation Institute.

1995. *Enhancing the Effectiveness of Modified Atmospheres to Control Insect Pests in Museums.* Getty Conservation Institute Scientific Program Internal Final Report. Los Angeles: Getty Conservation Institute.

**Soderstrom, E. L., D. G. Brandl, and B. E. Mackey**

1992. High temperature combined with carbon dioxide enriched or reduced oxygen atmosphere for control of *Tribolium castaneum* (Herbst) (Coleoptera: Tenebrionidae). *Journal of Stored Products Research* 28(4):235–38.

**Soderstrom, E. L., B. E. Mackey, and D. G. Brandl**

1986. Interactive effects of low-oxygen atmospheres, relative humidity, and temperature on mortality of two stored-product moths (Lepidoptera: Phyralidae). *Journal of Economic Entomology* 79(5):1303–6.

**Thermo Lignum Ltd.**

1995. Personal communication from Werner v. Rotberg, Heidelberg, Germany.

**Valentin, N.**

1993. Comparative analysis of insect control by nitrogen, argon and carbon dioxide in museum, archive and herbarium collections. *International Biodeterioration and Biodegradation* 32:263–78.

**Valentin, N., and F. Preusser**

1990. Nitrogen for biodeterioration control on museum collections. In *Biodeterioration Research*, vol. 3, 511–23. New York: Plenum Press.

Chapter 2

# Methods and Materials for the Anoxia of Insects in Museum Objects

After summarizing the practice of insect anoxic treatment, this chapter gives a brief rationale for the selection of the best type of enclosure or chamber for anoxia. It then deals extensively with the materials and tools required to construct bags or larger tents and with the equipment needed for anoxia in these containers as well as in commercially available chambers.

## Anoxic Eradication of Insects

Infested objects are placed in an airtight enclosure that can maintain an oxygen-free environment. The atmospheric oxygen in the enclosure is then either removed by placing oxygen absorber packets with the objects and sealing the enclosure or flushed out by passing through an externally supplied nitrogen gas. The required oxygen concentration of 3% by volume is maintained for the duration of the recommended fourteen-day treatment. Because the insect mortality in oxygen-free environments depends strongly on temperature and oxygen concentration, these parameters have to be monitored and well controlled during the treatment. Objects that are subject to insect infestation are prone to damage caused by large changes in environmental relative humidity. Therefore, the relative humidity of the treated object must also be maintained throughout the treatment.

### The containers and the method

The anoxic kill of insects in museum objects by oxygen deprivation can be performed in small to quite large plastic bags, in tents of 10 cubic meters or more in size, in commercial fumigation tents of 30 $m^3$ or more, or in stationary rigid chambers ranging from a volume of 2 to 3 to many cubic meters.

The basic anoxia technique often used in small or medium-sized plastic bags is to remove oxygen from the air by placing oxygen absorber packets in them (MGC 1994). It is impractical to use this method in large enclosures because each packet absorbs a limited amount of oxygen and is relatively expensive. Therefore, the air in large bags or tents (and sometimes in medium-sized bags) is displaced by passing nitrogen gas through them from cylinders or other nitrogen sources. The nitrogen flow is continued until the oxygen level decreases below 0.3%.

(From seven to eight times the volume of a bag is the minimum amount of nitrogen gas required to flush out enough air to achieve this low oxygen concentration.) Then, or before, packets of oxygen absorber are placed inside the bag to absorb any oxygen from air that might leak in during the anoxia treatment and the bag is sealed. With the correct number of packets of oxygen absorber in the bag and proper sealing, the oxygen content of the atmosphere in the bag will remain below 0.3% for two weeks or longer. The recommended minimum time for anoxia treatment in North America is fourteen days at 25°C. In Europe or regions where an anoxia-resistant insect, such as *Hylotrupes bajulus* (the old house borer), is indigenous, a twenty-two-day anoxia period is safest to ensure complete kill of all life stages.

### Safety of objects in nitrogen atmospheres

There may be concern about the effect of nearly pure (99.7%) nitrogen atmospheres on museum objects. Nitrogen is an inert gas; it is totally nonreactive with any substance in a collection of museum objects at any museum temperature. Researchers have long recommended the storage of artifacts in nitrogen or other inert gases to slow the oxidative and photochemical deterioration of sensitive organic materials (Kuehn 1968; Maekawa, Preusser, and Lambert 1992). Today important documents and objects, such as the Charters of Freedom of the United States, the Constitution of India, and the Royal Mummy collection in the Egyptian Museum of Cairo, are displayed or stored under a helium, argon, or nitrogen atmosphere (Calmes 1985; Maekawa, Preusser, and Lambert 1992).

It is true that the absence of oxygen—not the presence of nitrogen—causes a minority of dyes and pigments in watercolors to fade more rapidly in a low-oxygen environment than in air (Arney, Jacobs, and Newman 1979). However, the time required for this fading to become apparent is far beyond the fourteen to twenty-one days needed for insect eradication. Kenjo's (1980) report that sienna (which contains iron oxide) and litharge (a lead oxide) fade in low-oxygen conditions has been refuted by the more recent work of Koestler and coworkers (1993).

The only possible risk to artifacts during anoxia treatment is damage resulting from changes in their moisture content. Nitrogen from a cylinder or other external source contains only traces of moisture, and thus a drastic reduction in the relative humidity of an object's environment could occur if this near-dry nitrogen is flushed through an anoxic enclosure. However, any "RH shock" to an object can be easily avoided by humidifying the nitrogen before it enters the treatment enclosure. (This procedure is fully described in Chapter 4.)

However, most oxygen absorbers release water as they react with oxygen. Therefore, when moisture-sensitive objects are being subjected to anoxia in the presence of oxygen absorbers, humidity buffering material must be placed in the enclosure to keep the relative humidity constant. Conditioned silica gel is an effective agent for this purpose. (See Chapter 3 for details of the procedure for controlling relative humidity during small-scale anoxia.)

### Bases for choosing the best mode of anoxic treatment

Obviously, there are many factors to consider in proceeding to solve an insect infestation problem with anoxia.

- What is the magnitude of the immediate insect threat?
  - –Is the number of small objects to be treated a few dozen or many?
  - –If there are many, can they be placed on shelving in an enclosure?
  - –Are there a few or a large number of bulky objects?
- What is the forecast for future needs?
  - –How many of what size objects will need treatment per year?

An evaluation of this kind is a useful start. If only a few dozen objects are involved per month, the time and equipment cost can be modest. At the other extreme, large museums or ethnographic collections may require continued anoxic treatment. They could be served best by installation of a rigid chamber with automatic controls—a convenient but high-capital-cost solution. Thus economic constraints in the individual museum have to be balanced against the convenience or desirability of particular anoxia procedures.

The following sections describe enclosures or containers for anoxia that vary greatly in cost and labor-intensiveness in their construction and use. They are intended to aid the choice of an anoxia procedure that will conform best to the needs of many types of collections and economic situations. (See Chapter 5 for examples of insect infestation problems and solutions.)

### Small- versus large-scale containers

Small-scale anoxia treatment can be performed easily using individual plastic bags fabricated in-house or commercially available prefabricated bags. Their stepwise construction is described in Chapter 5. Oxygen-barrier film is moderately expensive and the fabrication of bags is time-consuming, especially if a large number of artifacts are to be treated. Therefore, in some situations the individual pouches and prefabricated tubes sold by conservation supply companies may be preferable.

Although bags can be reused, there are limitations in their durability and applicability. Tents and large flexible or rigid chambers are designed for reuse and allow treatment of a large number of objects at the same time, either by placing them on shelves or by stacking them carefully. However, the initial setup is more labor-intensive and far more costly than the preparation of a few dozen medium-sized bags. In their operation large tents or chambers require comparably greater amounts of nitrogen or an unreasonable quantity of oxygen absorber. Further, handling large amounts of oxygen absorber can be inconvenient as well as costly, and there is the possibility of the absorber generating object-damaging heat if it is placed directly on an object. Thus flushing air from large enclosures with nitrogen rather than absorbing oxygen from the air is the preferred method of achieving low-oxygen atmospheres inside them.

This use of nitrogen from external sources entails additional initial costs for flow regulators and RH monitors, as well as the fabrication of a humidifying system with flowmeters. Also, with nitrogen flow, closer monitoring of the oxygen content in the treatment enclosure is necessary so that only a minimal amount of nitrogen is used continuously or periodically to purge the enclosure and maintain its oxygen content below 0.3%.

In summary, plastic bags of small to medium size are preferable for the anoxia treatment of a few infested objects. Initial costs and operational costs—other than labor—are relatively low. Reuse of bags for anoxia is possible but limited. Larger tents or rigid chambers can treat many more objects at a time and can be permanent assets for future anoxia needs. However, because of their essential monitoring and gas delivery systems, they have higher initial capital costs. Labor charges for constructing large tents are also higher than those for even a sizable number of small to medium bags. Despite these factors, the advantages of the large tent must be considered: (1) capacity for treating many small objects at one time; (2) capability to disinfest bulky objects; (3) lower labor costs for operation per object treated along with comparable nitrogen versus oxygen absorber cost per cubic meter of anoxia volume; and (4) availability for future anoxia needs on short notice.

If budgets permit and forecasts of future needs warrant it, an obvious optimal solution is to have both very large and modest anoxia capability in-house. This translates to a large flexible (or rigid) enclosure with its permanent monitoring and nitrogen supply equipment always available, as well as with the expertise and simple tools to make plastic bags for special pieces or tasks. In the smaller bags an oxygen absorber can be the mode of removing oxygen. "Nitrogen supply equipment" here means the accessory apparatus needed for humidifying nitrogen in addition to having the gaseous or liquid nitrogen source itself (see Chapter 4). In most cities nitrogen gas cylinders or liquid nitrogen tanks can quickly be delivered to a treatment site.

A less costly alternative is to acquire expertise in bag production of small to medium or odd-shaped bags (e.g., for tapestries) along with valves, monitors, and so forth, so that either an external nitrogen supply or packets of oxygen absorber can be used. With this equipment on hand, one can rent or purchase a large commercial tent, or Bubble, for unusual needs. Obviously, rental reduces the fabrication costs (and permanent space required) for an in-house tent if the need for insect elimination is infrequent or a onetime event.

For many collections, it is most economical to use only oxygen absorber in moderate-sized bags for anoxic treatments (see Chapter 3). This approach can be successful in small or even large museums if the insect problem is minimal and if all suspect incoming objects are treated before installation. An occasional very large or odd-shaped object that needs to be treated can be handled as a special case. Fabrication of a bag to deal with its nonordinary configuration is not difficult. Admittedly, a nitrogen supply and its accessories must be used if the bag is much larger than about 300 L.

## Oxygen-Barrier Film for Making Bags and Tents

Flexible anoxia enclosures are readily fabricated from plastic film through which oxygen is not easily transmitted. Fortunately, the skill required to seal thin sheets of this type of plastic with a heat sealing device, and thereby form leak-proof pouches or bags, can be developed with a little practice.

### Overview

The properties of plastics that are sold in thin sheets or films vary enormously. There are many contributing factors, for example, the wide variation in the chemical nature of the individual molecules that join to form the polymer (the "plastic"); the number of such monomer molecules per polymer chain (the "molecular weight"); and the mode of fabrication of the finished plastic film. The flexibility of the final polymeric film is often improved by adding plasticizers, its resistance to light or oxygen (and thus to "aging") by adding stabilizers. The simultaneous polymerization of two or more monomers yields a copolymer whose properties are sometimes superior to either monomer's individual polymer. Desirable film characteristics, such as strength or resistance to gas or vapor permeability, can be improved through special treatments of the finished polymer film. These include stretching while being heated, a process that causes the polymer chains to orient themselves in more orderly molecular arrangements and greatly increases their tensile strength.[1]

However, no physical or chemical modification of a polymeric film can produce one that is perfect in all respects. Rather, to obtain a film that has a number of disparate desirable characteristics, the approach is to combine layers of quite different films to form a composite in a laminate form. The commercial process involves heating and pressing several layers of film, each selected for a particular superior property, such as strength, durability, or impermeability to a specific gas, into a single thin pliable film.

Plastic films that are optimal for the fabrication of anoxia bags or tents should be flexible (rather than brittle or too stiff), easy to seal to one another (a significantly lower melting temperature at the sealing layer compared to nonsealing layers), resistant to distortion while being heat sealed, tough (not easily punctured or scratched), strong (stretched or torn only with difficulty), oxygen impermeable (free of voids, not a good solvent for oxygen), transparent (so that an object or a monitor can be seen through it), and inexpensive. This list is as daunting to achieve technically as it is long. Nevertheless, modern composite films can meet these requirements at the high cost of $5 to $11 per square meter. Only aluminized—and thus opaque—laminated films are inexpensive, in the range of $2 per square meter.

If a generic oxygen-barrier composite film is examined from the inside to its outer surface, it would consist of three to five layers

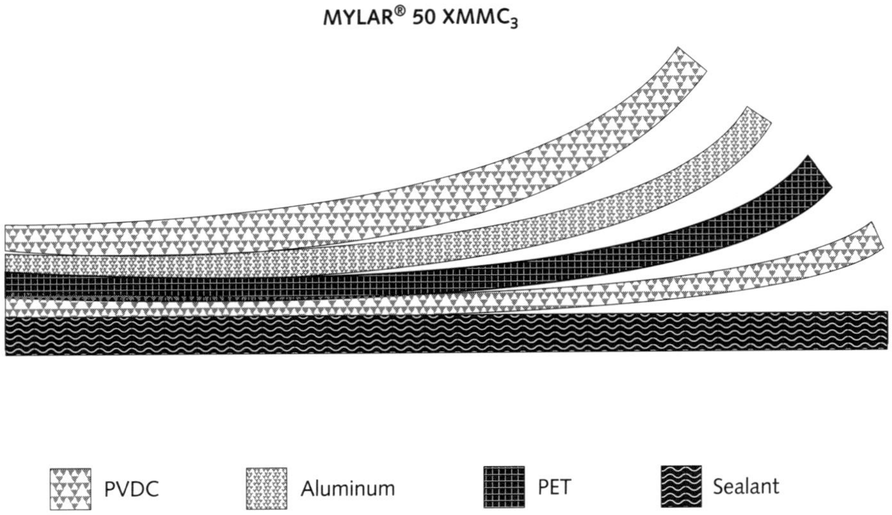


**Figure 2.1**
Composition of a barrier film laminate (Dupont 1994)

(Fig. 2.1). The inner layer is the sealing film that softens to near-melting at a relatively low temperature so that when it is pressed against a comparable hot film, the two will dissolve and thereby unite in a perfect seal. The next layer is often the oxygen-barrier film, extremely impermeable to molecules of oxygen. Then a layer on top of that oxygen barrier must be present to prevent any damage to the barrier film that would disrupt it and allow oxygen to flow through. This top layer must be a tough, puncture- and tear-resistant polymeric film such as Mylar, a type of polyethylene terephthalate (PET), or nylon. Some films have a moisture barrier layer to improve the oxygen barrier's performance in moist environments.

A list of polymers and chemicals and their corresponding abbreviations and acronyms is provided in Table 2.1. Here, to conform to common usage rather than to modern chemical nomenclature, the somewhat confusing parentheses are omitted. For example, rather than the precise usage *poly(ethylene)*, we use the common form *polyethylene*.

A fourth layer of the film might be another tear-resistant layer or an extremely thin aluminum coating (formed by vaporization of aluminum, not conventional aluminum "foil") that is almost completely impervious to oxygen. However, this kind of aluminum coating is subject to minute pinholes or voids and easily damaged. Therefore, another protective film layer should be pressed on top of the aluminum to protect its integrity. Whether five layers or three, the

**Table 2.1**
Abbreviations and acronyms for polymers and chemicals

| | |
|---|---|
| Acrylonitrile butadiene styrene copolymer | ABS |
| Cellulose acetate | CA |
| Cellulose acetate butyrate | CAB |
| Chlorinated poly(vinyl chloride) | CPVC |
| Ethylene chlorotrifluoroethylene copolymer | ECTFE |
| Ethylene methyl acrylate copolymer | EMAC |
| Ethylene tetrafluoroethylene copolymer | ETFE |
| Ethylene vinyl alcohol copolymer | EVOH |
| Ethylene vinyl acetate copolymer | EVA |
| Ethylene vinyl chloride copolymer | EVCL |
| High-density poly(ethylene) | HDPE |
| Low-density poly(ethylene) | LDPE |
| Linear low-density poly(ethylene) | LLDPE |
| Medium-density poly(ethylene) | MDPE |
| Methyl ethyl ketone | MEK |
| Oriented poly(propylene) | OPP |
| Poly(acrylonitrile) | PAN |
| Poly(butylene terephthalate) | PBT |
| Polycarbonate | PC |
| Poly(chlorotrifluoroethylene) | CTFE or PCTFE |
| Poly(ethylene) | PE |
| Poly(ethylene terephthalate) | PET |
| Poly(methyl methacrylate) | PMMA |
| Poly(propylene) | PP |
| Poly(styrene) | PS |
| Poly(tetrafluoroethylene) | PTFE |
| Polyurethane | PU |
| Poly(vinyl acetate) | PVAc |
| Poly(vinyl alcohol) | PVAL |
| Poly(vinyl butyral) | PVB |
| Poly(vinyl chloride) | PVC |
| Poly(vinylidene chloride) | PVDC |
| Poly(vinylidene fluoride) | PVDF |
| Vinyl acetate | VA |
| Vinyl acetate ethylene copolymer | VAE |
| Vinylidene chloride | VDC, VdC |

goal is an oxygen-impermeable film that is easy to handle, tough, and capable of being made into bags.

The final oxygen-barrier laminate seals easily—inner layer to inner layer—because the hot sealing iron is not "gummed up" by direct contact with it: the iron presses only on outer layers, which have much higher melting temperatures, not directly on the inner low-melting, low-density polyethylene or similar plastic.

### Some laminates useful for anoxia and a comparison of their properties

Listed below are commercially available laminated oxygen-barrier films that are desirable for anoxia enclosures. They are ordered according to their oxygen transmission rate (OTR), or the rate at which oxygen can pass through the film, an important technical measurement that is discussed below. (This is not the only criterion that should be considered.) Toughness, maintenance of the oxygen barrier during handling, and ease of sealing are also vital and are discussed below (pp. 24–27). OTR values of monolayer polymers are given in Table 2.2, and laminate properties are given in Table 2.3. (See Appendix IV for suppliers.)

#### *Aluminum foil. Aluminized film: Laminate OTR of 0.01 to 0.05*

Oxygen does not pass through aluminum or other metal foil. Clearly it is a perfect oxygen barrier. However, it is difficult to seal foil to other polymers in a laminate that can then be used to fabricate bags. A compromise is aluminized film, prepared by vaporizing aluminum and depositing it as an extremely thin coating on various kinds of polymers. Unfortunately, this vaporized coating cannot be flawless and inevitably there are occasional pinholes or small voids. The oxygen-barrier qualities are excellent, as shown by the data in the heading of this paragraph. However, serious flaws that allow oxygen to pass can develop with folding and with wear if the coating is not protected by an additional tough laminate layer or

**Table 2.2**
OTR values and heat-seal ranges of nonlaminated polymers

| Polymer Film, 0.025 mm Thick | OTR, cc/m²/day, Tested at 23° to 25°C, 0% RH | Heat-Seal Range, °C |
|---|---|---|
| PVDC | 2.4[a] | 120–150[b] |
| EVOH | 12[a] (at 100% RH) | NA |
| CTFE | 47[c] | 199–227[d] |
| Nylon | 30–110[d] | 177–260[d] |
| PET | 132–142[e] | 250–265[d] |
| | 56–108[a] | melting point |
| PVC | 150–350[f] | 93–176[b] |
| HDPE | 520–3900[g] | 135–155[g] |
| OPP | 2000–2500[f] | NA |
| PP, cast | 3700[f] | 140–205[b] |
| LDPE | 3900–13,000[g] | 121–177[g] |
| EVA | 7200[h] | < 120[f] |
| Surlyn® | 5200–11,800[h] | 130–163[i] |

[a] Corish 1992. [d] Paine 1991. [g] Sweeting 1971b.
[b] Goodfellow 2000. [e] Benning 1983. [h] Plastics Design Library 1995.
[c] Finlayson 1989. [f] Briston 1983. [i] Cryovac 1996.

**Table 2.3** Properties of laminated films (plus HDPE and PE monolayers)

| Trade Name | (Manufacturer) or Supplier | Composition | Thickness (mm) | (mil) | OTR ($cc/m^2/day$) Spec. | GCI test | Heat-Seal Temperature °C | Costs[2] ($\$$U.S./$m^2$) |
|---|---|---|---|---|---|---|---|---|
| ESCAL™ | (Mitsubishi Gas Chemical America Co., New York, NY 10022) [MGC] | (SiO) × barrier Additional layers unknown | 0.11 | 4.2 | 0.05 | NA | NA | 8.00–10.00 also available as bags |
| Filmpak 1193 | (Ludlow Corp., Homer, LA 71040) | Polyester PE Probably PVDC PE | 0.13 | 4.9 | 0.1 | 0.28 | NA | 5.30–8.50 |
| PTS | (Mitsubishi Gas Chemical America Co., New York, NY 10022) [MGC] | NA | 0.11 | 4.2 | 0.5 | NA | NA | 7.00– 9.00 available as bags |
| CAL DRY 320 | Cal Tex Plastics, Vernon, CA 90058 | Polyolefin Polyester | 0.1 | 4 | < 8.47 | 1.05 | NA | 4.00 |
| ArchiPress Pouches | Multipak B.V., (3881 LB Putten, Holland) | Polyester Additional layers unknown | 0.092 | 3.6 | 6 | 2.3 | 120–145 | 7.00–12.60 available as bags |
| Cryovac BDF-2001 | (Cryovac Division Sealed Air Corp., Duncan, SC 29334) | LDPE/MDPE EVA Adhesive layer EVOH | 0.03 | 1 | 4 | 3.72 | NA | |
| PET/Saran/ PE Film | Keepsafe Systems, Inc., Toronto, Canada | Polyester PVDC PE | 0.06 | 2.5 | NA | 7.1 | NA | 4.40–5.50 available as bags |
| Saranex 23 | (Dow Plastics, Midland, MI 48674) | LDPE EVA PVDC EVA | 0.05 | 2 | 7.75 | 7.5 | 115.5–221 | |
| Aclar, Bell Fibre FR-7750, Filmpak 1177 | (Honeywell, Inc., Morristown, NJ 07960) | CTFE (Aclar) LDPE Polyester LDPE | 0.11 | 4.2 | 51.15 calcu-lated value | 50.27 | 190.5 | 6.00–11.00 |
| Marvelseal | (Ludlow Corp. Homer, LA 71040) | Nylon PE Aluminum foil PE | 0.14 | 5.3 | 0.01 | 0.01 | 204.4 | 1.70–2.00 |
| Shield Pack Class A | (Shield Pack, Inc., West Monroe, LA 71292) | OPP PE Aluminum PE tie layer LLDPE | 0.13 | 4.9 | < 0.14 | 0.01 | NA | |
| Fumaplas M12 | (Power Plastics Thirsk, North Yorkshire, UK, YO7 1PZ) | HDPE Polyester Aluminum I | 0.11 | 4.2 | 0.05 | NA | NA | Sold in form of CAT Bubble |
| Valsem® | (SNEC, 60190 Lachelle, France) | Aluminum I Additional layers unknown | 0.14 | 5.3 | NA | NA | NA | |
| HDPE Film Sheeting | McMaster-Carr, Los Angeles, CA 90054 | Cross laminated HDPE | 0.1 | 4 | NA | 567 | NA | 1.10–1.20 |
| Shannon Shield 1115 | (Shannon Packing Co., Chino, CA 91710) | Polyester Vacuum-deposited Aluminum Black polyolefin | 0.09 | 3.5 | 0.008 | NA | NA | |
| Poly Tubing | (National Bag, Hudson, OH 44236) | PE | 0.1 | 4 | NA | 2583 | NA | 0.96 |

handled with extreme care. Commercial examples are Marvelseal 360, Shield Pack Class A, Fumaplas M12, and Valsem®.

*Polyethylene terephthalate (PET): Laminate OTR of 2.3*

PET is a polyester. It is relatively inert and has a high tensile strength as well as a high softening point. (Fibers that are extruded from molten PET are stretched to orient them and spun into Dacron. When the molten PET is cooled in thin sheets, it forms the film pioneered by DuPont, Mylar.) Its gas barrier properties are good to excellent. Also, it is a good barrier to water vapor and has extremely high resistance to puncture and tearing. Therefore, it is used as a strengthening or toughening layer in addition to being an oxygen barrier in a laminate. As a composite with aluminum, its strength yields an unbeatable oxygen barrier that is preferred for very large anoxia tents, the commercial Bubbles. ArchiPress Pouches (OTR, 2.3), Fumaplas M12 (OTR, 0.05).

Some PET foils and other laminates can be obtained without heat-seal layers. These can be sealed with high-temperature impulse sealers described in the next section.

*Polyvinylidene chloride (PVDC): Laminate OTR of 0.3 to 8*

PVDC is an outstanding barrier to gases and water vapor, and its properties are almost unaffected by changes in moisture content. Heat-sealable layers often consist of copolymers of vinylidene chloride with vinyl chloride; their low permeability to gases and water vapor augments the excellent oxygen-barrier properties of the PVDC layer in the laminate (Briston 1983). Saran produced by the Dow Chemical Company is the common trade name for PVDC. Filmpack 1193 (OTR of 0.28), Cal Dry 320 (OTR, 1.0), Cryovac BDF-2001 (OTR, 3.7), KeepSafe (OTR, 7.1), Saranex 23 (OTR, 7.5).

*Copolymers of Ethylene–Vinyl Acetate (EVA) and Ethylene–Vinyl Alcohol (EVOH): Laminate OTR of 3.7*

The copolymer EVA is a top-performing gas barrier and can be heat sealed. EVOH has similar properties, but it is hygroscopic (Sweeting 1971a). Thus its ability to act as an oxygen barrier decreases drastically at high humidity levels. It is still a good heat-seal layer in a laminate, as are other copolymers such as Surlyn®, an ethylene–methacrylic acid copolymer (Paine 1991). Cryovac BDF-2001 (OTR, 3.7).

*Polychlorotrifluoroethylene (CTFE): Laminate OTR of 50*

CTFE, manufactured by Allied Signal, Inc., and sold under the brand name Aclar, has the lowest water vapor permeability of all polymer films. It is not as good in terms of restricting the movement of gases through it but is still a fair barrier against oxygen transmission (Paine 1991). The toughness of the laminate commends it for use in anoxia bag construction. Filmpak 1177, FR-7750, Film-O-Rap 7750 (OTR, 50).

Anyone who works with the preceding laminates will encounter other polymer films. Below we describe some of these and explain reasons for their use, as well as why they are *not* recommended as oxygen barriers in anoxia bags.

**Polyethylene** (PE), a chemically inert material, is unsuitable as a barrier film for anoxia enclosures because of its relatively high permeability to oxygen (Benning 1983), some 10 to 300 times worse than any recommended anoxia composite plastic. However, it is commonly used in nearly all oxygen-barrier laminates as the heat-sealing layer. The abbreviations LDPE and HDPE stand for low- and high-density polyethylene, respectively. Polymerized by different chemical processes, they differ greatly in their physical properties. LDPE is softer and melts at a lower temperature over a wider range than HDPE. HDPE, the more linear PE, is stronger and tougher than LDPE and is present in several barrier laminates as the outer protective layer.

**Polypropylene** (PP) is stiffer than polyethylene. Cast polypropylene is not as "porous" to oxygen as the worst LDPE, but it is still unusable as an oxygen barrier even when it is improved by a stretching process[2] during formation of the film. However, the resulting molecularly oriented polypropylene (OPP) is stronger and tougher than conventional PP. Modified OPP is used as a heat-sealing layer (Benning 1983; Sweeting 1971b). Because their monomers were once referred to as "olefins," PP and PE are sometimes called polyolefins.

**Nylon** is the generic term for several different kinds of polyamides developed by DuPont. Properties differ for the various nylons, but in general they are strong, have a high tensile strength, and are slightly hygroscopic. They are not very good barriers to water vapor but are good gas barriers (Briston 1983). However, the permeability for oxygen increases with increasing relative humidity (Finlayson 1989). Nylon can be laminated with PE and EVOH or coated with aluminum to produce heat-sealable laminates with a very low oxygen permeability.

**Polyvinyl chloride** (PVC) is hard and fairly brittle, but it can be modified by the addition of plasticizers to make it softer and more flexible (Briston 1983) for many everyday uses, from garden hoses to shower curtains. Unplasticized PVC is a relatively good barrier to oxygen, but the barrier properties are greatly reduced by the incorporation of plasticizers (Paine 1991). The action of light and heat causes a very slow liberation of hydrogen chloride, with the consequent degradation of the polymer (Sweeting 1971b), making it unsuitable for use with valuable artifacts for longer than a few weeks.

Table 2.2 summarizes oxygen barrier values (OTR) and heat-seal temperature ranges for many of the polymers described above. Similarly, Table 2.3 lists those characteristics for a number of laminates, with their approximate costs. Details of the standard test method for the OTR of films are given in the next section.

## The selection of a barrier film for an anoxic bag

The choice of a barrier film is not difficult. Any one of the laminates described in the preceding section would adequately restrict the transmission of oxygen through the walls of an anoxia enclosure it forms. Each can readily be heat sealed to make such an enclosure. The only questions to be decided are cost versus labor time, small bag versus large tent, and where to buy the barrier film.

Aluminized laminates are the least expensive per square meter but, except for the claimed superiority for Fumaplas M12 and Valsem®, they develop pinholes relatively readily with use. Further, they are opaque. It is a decided advantage to be able to observe oxygen and relative humidity indicators or monitors or other devices that may be placed in a bag. Beyond this, it is an advantage to view at any time the object being treated. This obstacle of an opaque laminate can be overcome by sealing a "window" of transparent laminate in the aluminized bag. However, until an individual develops considerable expertise, window installation requires considerable time and effort (see Protocol A). Furthermore, any additional length of seal, such as the four sides of a window, increases the possibility of leakage. In other words, the cost saved by using a barrier film that is an aluminized laminate can be more than offset by additional labor requirements and increased risk.

The choice between laminates is rather easy when it is only a matter of which to use for making small or medium bags versus which is best for fabricating a much larger reusable tent. All the listed laminated films are suitable for smaller anoxia enclosures. However, no laminate that is less than 0.1 mm (4 mil) thick—even with a good OTR rating—should be considered for a large tent. The thicker, more rugged film is essential for mounting fittings through it, for handling such a big unit, and for its integrity on reuse.

It is unfortunate that the absence of a local or a not-too-distant supplier of plastic film may severely limit which laminate one can adopt for anoxia. Although a generous number of suitable laminates have been described here, all are not easily purchased in small quantity. Manufacturers will not deal with museums or individuals who want to buy less than thousands of dollars of an item. Suppliers in large cities that purchase from manufacturers for reselling in smaller lots often have a minimum order of at least $100 and rarely stock more than two kinds of barrier films. A few resellers will provide small samples of laminates so that an individual can test its sealability or run an Oddy test for its potential for harming museum metals. (A personal visit to the company is usually more effective than impersonal telephone calls in getting samples for tests. The seller can become interested in why a museum would want to use special plastic package wrap.)

A good source of several but not all of the best laminated oxygen-barrier films is a conservation supply company. Each of those listed in Appendix IV provides good service, but none carries a wide range of laminates. Obtaining a particular film locally may not be easy if you do not live in a large city. Therefore, it is helpful that a very adequate oxygen-barrier film can be obtained even from a distant seller of conservation materials.

Two technical notes concerning barrier films that may be of interest are the safety of objects in terms of their contact with or nearness to plastic films and the validity of measurements of the flow of oxygen through the laminates (the OTR) described in this section.

First, with regard to hazards to museum objects posed by chemicals present in plastic oxygen-barrier films: the polymers that are used in the laminates are not a threat at normal temperatures; they are stable substances that do not change or produce noxious materials under museum conditions. However, plasticizers—chemicals that are used to increase the flexibility of some kinds of polymers—may be present in some laminates in small quantities. There have been no reports of damage to any object caused by them in thousands of anoxia applications. If the conservator wishes even more assurance than this extensive experience provides, an Oddy Test of twenty-one days would be helpful (Green and Thickett 1995) using small samples of potential barrier films. The simple step of using tissue to wrap objects that might come in contact with a film would avoid any hazard. Alternatively, placing objects in cardboard boxes (open on one side for easy observation of the contents) would prevent them from coming into contact with the film.

The oxygen transmission rate of plastic films is measured by the standard method of the American Society for Testing Materials (ASTM 1999) wherein the film sample is mounted as the dividing barrier between two small chambers. Then pure nitrogen flows continuously through one chamber while oxygen flows through the other. Any oxygen permeating through the film into the nitrogen chamber is transported to a coulometric detector, producing an electrical current that is proportional to the amount of oxygen flowing into the detector per unit time.

The ASTM oxygen transmission rate is measured by the volume of oxygen that moves through a sample of a film per square meter of film area each day ($cm^3/m^2$/day), at 20°C to 25°C and 0% RH at the normal atmospheric pressure. (OTR values in a table apply only to the thickness of the polymer film that is listed in that table.)

Unfortunately, a few manufacturers include U.S. dimensions in the standard ASTM OTR values. Rather than the ASTM units of $cm^3/m^2$/day, they state OTR results in $cm^3$/100 $in.^2$/day. The conversion factor is simple: ASTM OTR = 15.5 U.S. System OTR for a film whose thickness is listed in a table.

In contrast, the *permeability* of a film is defined as dependent on the *unit* thickness of the film. (Thus it technically cannot be applied to a laminate. Laminates are not homogeneous materials, and their properties are not necessarily uniform with an increase in thickness.) Nevertheless, if a film's permeability is multiplied by its thickness in mm or mil (0.001 in., i.e., 0.025 mm), the result is approximately the OTR. (The conversion factor is convenient between the ASTM permeability of ($cm^3$·mm(thickness)/$m^2$/day) and the U.S. System permeability of ($cm^3$·mil(thickness)/100 $in.^2$/day): 2.54 ASTM permeability = U.S. System permeability.) See Appendix III for a table of conversion factors.

Often the OTR and test conditions are available from manufacturers in a data sheet for their product (Fig. 2.2). The figures provided by the various manufacturers are sometimes confusing because the OTR is determined with air rather than 100% oxygen. Fortunately, the comparison is easy: simply dividing an OTR value in pure oxygen by 4.8 (because air is only 20.9% oxygen) gives its OTR in air. Films with an oxygen transmission below 20 cc/$m^2$/day, tested at 100% oxygen concentration (or in air, about 4 cc/$m^2$/day) are preferable for anoxia work.

**data sheet**

Ludlow Coated Products

Homer, LA 71040  Meridian, MS 39301  Columbus, GA 31906

04/25/97

DATA SHEET
for
Filmpak 1193

MIL-B-22191D, Amendment 1, TYPE I, CLASS 2

| | | | | |
|---|---|---|---|---|
| WVTR | 0.03 | g/100 in/24h @90%RH,100 F | Nom | ASTM F 1249 |
| Oxygen TR | .05 | cc/m /24h @0%RH,73 F | Nom | ASTM D 3985 |
| COF (hs:steel) MD | Static 0.24 | Kinetic 0.19 | Nom | ASTM D 1894 |
| Thickness | 4.9 | Inches | Nom | TAPPI T411 |
| TEA MD/CD | 170/35 | ft-lb/ft2 | Nom | TAPPI T494 |
| Dyne Level | 35 | dynes | Max | TAPPI T698 |
| Tensile MD/CD | 30/22 | #/inch | Nom | TAPPI T494 |
| Puncture Resistance | 17 | lb. | Nom | FED101,2065 |
| Seam Strength | 10 | #/inch | Min | 350,50,.5 |

4904 square inches/pound +/- 10%

The reported values are representative of the product named. They do not encompass normal variability, or constitute any warranty.

24-1-042J

A tyco INTERNATIONAL LTD. COMPANY

**Figure 2.2**

Barrier film data sheet (Ludlow Corporation 1997)

Nevertheless, the CTFE (Aclar) laminate with its OTR of 50 (in oxygen) is suitable for small or medium bags for fourteen-day anoxia with the proper amount of oxygen absorber to react with any oxygen leaking in.

Table 2.3 (p. 22) summarizes information on the construction, thickness, oxygen transmission rate, and heat-seal temperature range as well as costs of several barrier films. The table is especially valuable because all OTR values in the "GCI Test" column were carried out under identical conditions at the Getty Conservation Institute rather than by individual manufacturers.

## Heat Sealing of Barrier Films to Form Anoxic Enclosures

In principle, making a bag from plastic film by folding a long rectangular sheet and heat sealing the edges with something like a thin hot iron is easy. As might be predicted, the actual process is not easy. Nevertheless, it is nearly so after an hour or more of practice if one is adept and chooses the right components for the job, the film, and the sealer instrument.

The choice of film is essentially already taken care of: barrier-film laminates are made to be readily sealed because their inner layer is a low-temperature sealant film, such as LDPE. Then the choice must be made between three types of tools to seal them: hand-held tacking iron (Fig. 2.3, $30–$80), hand-held bar sealer (Fig. 2.4, $100–$200), or bench-mounted impulse sealer (Fig. 2.5, $160–$1,000) that makes about a 2 mm wide seal for a length of 15 cm to 60 cm. Generally, it is easier to use a small hand-held tacking iron and hand-held bar sealer when fabricating tents and complex shapes of enclosures from barrier films. Bar sealers provide heat from two sides of the sealing films; therefore, they can provide a higher sealing temperature at sealing surfaces. A fixed or bench-mounted impulse heat sealer is very useful and efficient for making large numbers of identical small pouches and when thin films, which are easily stretched or deformed by heat, are used. (Both sealing and cooling times can be set on an impulse sealer. However, bench-mounted units are impractical to use on the final seal after an object has been inserted in the bag. There are many elaborate varieties of impulse sealers for double seals and for mass production.) It is a fair but not a hard-and-fast rule of thumb that hand-held bar sealers can be used for most tasks in fabricating tents, with the exception of complex in-the-corner joints. The temperature can

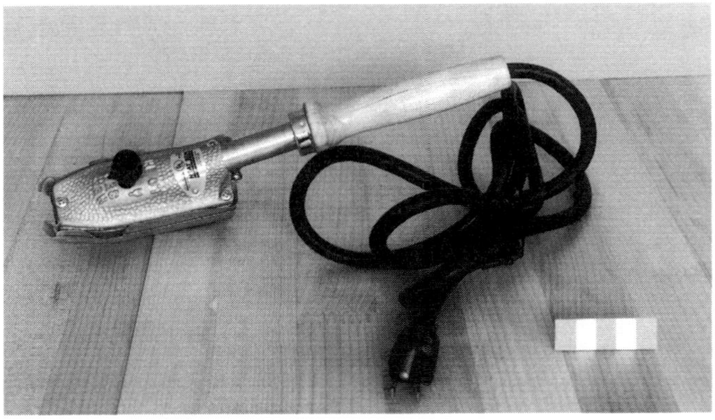

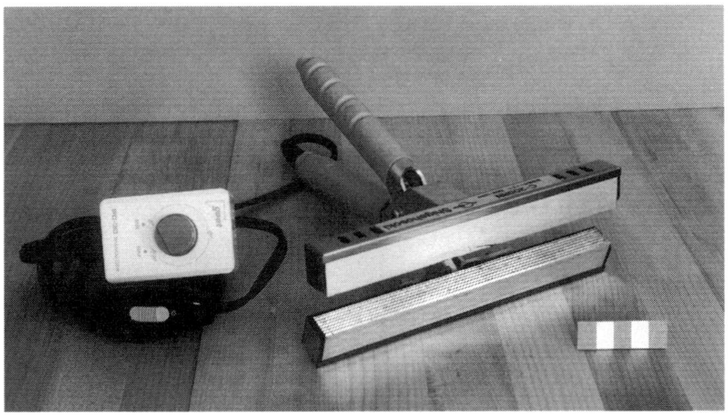

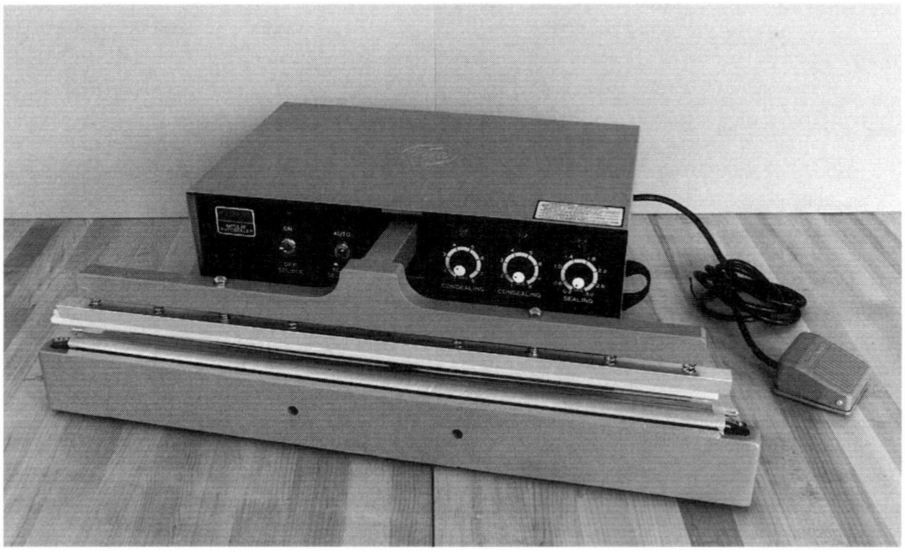

**Figure 2.3, above**
Tacking iron with adjustable temperature (scale = 5 cm)

**Figure 2.4, above right**
Constant temperature portable bar sealer (scale = 5 cm)

**Figure 2.5, right**
Bench-mounted impulse heat sealer

be adjusted in each type of sealer mentioned, and discovery of the best temperature for the laminate as used by each operator is a major part of the learning curve for sealing barrier film perfectly.

The skill of the individual operator is not a trivial consideration, primarily with regard to the use of a tack sealer. Polymer laminates differ in their heat transmission from the outside layer to the inner sealing layer. Therefore, the pressure exerted by the user and the steady rate at which the tacking iron is moved—highly individual factors—influence the quality of the final seal as much as the correct temperature in the sealer. The latter must not be hot enough to soften the top layer. Although all better sealers are Teflon coated, if the top layer approaches melting, it will still impede the steady movement of the iron and make less probable a smooth seal of the inner layers. An important hint is to place a paper towel on the hard smooth surface on which the laminate is resting as it is being sealed (Fig. 2.6) if tacking irons are used. This slight softness aids in making a superior closure.

For the preceding reasons, impulse and bar sealers require less training for operation than does a tack sealer. Once the optimal temperature and time of clamping the two jaws on a given laminate is found, even a novice can rapidly move a bar sealer along a designated seal line. A second "bulwark" seal line is readily made and constitutes insurance against leaks.

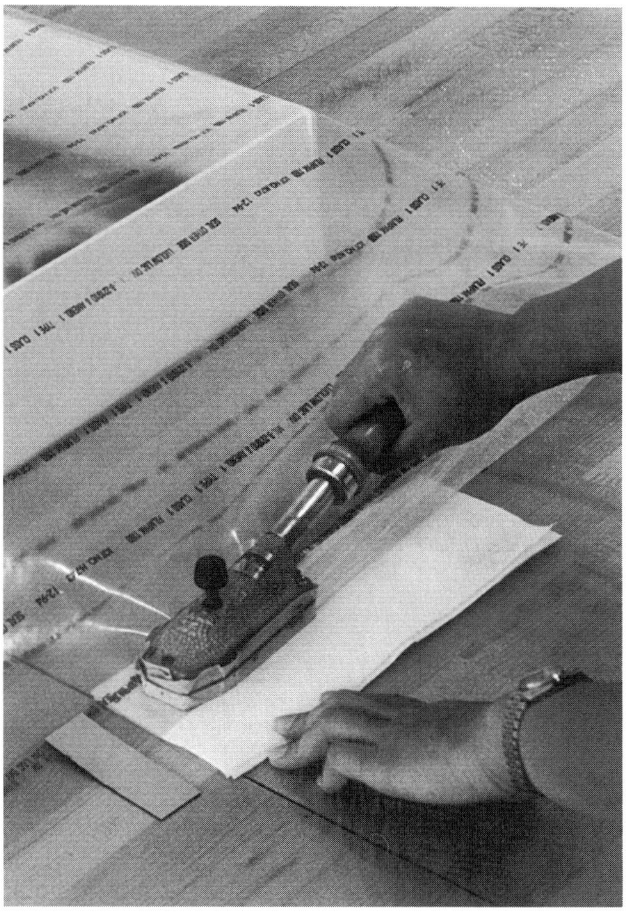

Figure 2.6
Plastic sheets are held in place by adhesive tape to maintain alignment; paper towel beneath heat-sealing iron aids seal.

When sealing big sheets of film together, it is convenient to use adhesive tape (see Fig. 2.6) and clips to hold them precisely in place before sealing. "Spot sealing," or tacking them at intervals with a hot iron, may seem like an obvious and a better way to keep them properly oriented. This is extremely inadvisable because almost inevitably both film sheets shrink or stretch unequally very slightly when heat is applied at spots. The wrinkles that form in the projected seam line are virtually impossible to press hard enough and heat thoroughly enough to prevent leaks at the wrinkle point. Similarly, it requires a higher level of skill to seal over a gusset or similar fold. The reason is that there is simply no free-flowing liquid to ooze into voids and fill them without seriously deforming other layers of the film. Hot melted polyethylene or other sealing polymer is not such a liquid. It is very soft when hot and in a melting range—one layer will dissolve in another; this is how two sheets are sealed—but it does not flow freely to fill holes and close gaps.

A few practical notes follow. Pillow or cubical shapes for bags and tents are preferable to complex shapes that follow the profile of a large object, because straight sheets are easier to seal together and the formation of leaky seams less likely. Experience at the GCI has shown that the use of ordinary clamps or sticky tape to seal bags as advocated by some conservators does not provide airtight seals. Conservators invest too much thought and time in an anoxia procedure to risk a leaking enclosure. Specially designed clamps for specific types of plastic pouches, when used on thicker plastic films than the ones for which they were originally designed, may not seal properly. Household bags with incorporated plastic zippers cannot be recommended for any type of museum anoxia because all those that were tested failed to seal. In contrast, larger, specially designed gas-tight zippers are available that provide a good seal. These zippers require an application of grease or a small amount of mineral spirits for gas-tightness. They are currently being employed in commercial reusable fumigation Bubbles that are described in Chapter 4.

Although there is no place in anoxic procedures for sealing everyday polyethylene or polypropylene film, conservators may want to make bags or similar nonbarrier enclosures to keep objects dust-free. These single-layer (nonlaminated) films cannot be sealed well with a tacking iron. They shrink and deform drastically as they are heated by the iron to their softening temperature range, making a neat joining impossible. For an adequate seal in polyethylene, an impulse sealer is necessary. Other types of polymeric films that are not used alone in anoxia, such as vinyls, Saran, Mylar, and cellulose acetate, may require electronic or ultrasonic welding devices for proper seals (Sweeting 1971b).

## Monitoring of Anoxic Conditions

The monitoring of temperature and relative humidity over the course of an anoxic treatment is recommended to ensure the safety of the infested museum objects. Monitoring the oxygen content in a treatment enclosure is highly recommended to guarantee complete eradication of insects. (Even in a carefully made enclosure, without monitoring one cannot tell if there is a small site where air is slowly leaking in and raising the oxygen level so that some insects can survive and later multiply.) Thus in this section we deal first with a simple visual oxygen indicator, then with electronic oxygen monitors. We discuss simple and complex monitors for relativity humidity and for temperature. The section concludes with a discussion of leak detectors and a brief introduction to the value of sensors connected to data acquisition systems such as dataloggers for documentation.

### Inexpensive qualitative oxygen indicators: Ageless-Eye®

Ageless-Eye® K, an inexpensive indicator for the presence of oxygen, is a small pink tablet in a blister packet that is sold in a package with Ageless® oxygen absorber so that it is exposed to less than 0.1% oxygen. Ageless-Eye® K will change from pink to violet within five minutes when the oxygen in the atmosphere around it rises to 0.5% or more (at approximately 25°C and 50% RH; Plate 1).[3] This behavior is reversible for many cycles. However, the process of changing back from violet in 0.5% to 20% oxygen to pink in about 0.1% oxygen is far slower, taking two to three hours (Plate 2). The color intensity of Ageless-Eye® K can fade if it is exposed to sunlight even for a short time or to fluorescent lights for long periods. When used inside an anoxia treatment bag, Ageless-Eye® K should be taped where it can be seen in the bag but shaded from bright direct light with a bit of aluminum foil or other opaque material.

Ageless-Eye® indicators come in various types suitable for different temperature and humidity ranges and slightly different oxygen levels. A summary of their properties is given in Table 2.4. Mitsubishi Gas Chemical Company (MGC) developed its Ageless-Eye® products using methylene blue, an organic dye, as the oxidation-reduction indicator (MGC 1997). The company states that it should not be reused and should be stored in a dark, cool, and oxygen-free atmosphere. This is easily accomplished by sealing the indicators accompanied by Ageless® oxygen absorbers in barrier-film bags that are then stored in a conventional refrigerator. Even under these conditions the guaranteed shelf life is only six months.

In the past, conservators reported problems regarding the reliability of the indicator's color change. For example, Gilberg (1989) described the poor oxidation-reduction characteristics of methylene blue at low relative humidity. He found that it failed to turn pink at 0% RH even after long periods in an atmosphere with less than 0.1% oxygen. He also reported that extended exposure to light caused the indicator to fail.

**Table 2.4**

Types of Ageless-Eye® oxygen indicators

| Ageless-Eye® Type/ Color of Blister Packet | Application Range (Temperature °C and % RH) | Color Change ppm Oxygen | Storage Recommendation/ Shelf Life |
|---|---|---|---|
| Type S*<br>Dry environment<br>String type | 5°–35°C<br>0%–92% | 1000–3000 ppm | Dark, cold, oxygen-free/6 months |
| Type C**<br>Standard type<br>White blister packet | 5°–35°C<br>30%–92% | 1000–1500 ppm | Dark, cold, oxygen-free/6 months |
| Type K<br>Blue blister packet | 15°–35°C<br>60%–92% | 2500–3000 ppm | Dark, cold, oxygen-free/6 months |
| Type A<br>Green blister packet | Replaced by Type C | | |
| Type T<br>Brown blister packet | Low temperature | | |

* Type S (string type) is used with RP-A oxygen and moisture absorber.
** Type C (blister packet without printing) is used in Ageless-Eye® master bags for quality assurance.

In extensive laboratory tests at the GCI, the performance of several batches of different types of Ageless-Eye®, C and K, was evaluated as to the effect of light, relative humidity, prolonged presence of air, reusability, and long-term use. Extended exposure to light was found not to affect the indicator's basic ability to change color in either direction. However, light caused so much fading—especially in type C—that determination of the color change was too subtle for practical use. Therefore, Ageless-Eye® should be stored in the dark and exposed to as little light as practical in anoxia work.

Ageless-Eye® type K showed a reliable color change from violet to pink when exposed to humidity levels between 50 and 100% RH at about 0.15% to 0.2% oxygen concentration, but its change was not consistent at lower RH values. Type C also responded very inconsistently when exposed to similar low humidity conditions and generally required oxygen concentrations below 0.01% to change to a pink color. For both Ageless-Eye® types, it was more difficult to identify a color change when tested at 100% RH because the tablets absorbed moisture and became darker. The conclusions are that Ageless-Eye® K and C are not useful near 100% RH or at RH levels below about 45%. Ageless-Eye® K is the superior type for use by conservators in anoxia procedures because it changes color in the 0.1% to 0.3% range. (This does not mean that it is superior to other brands of indicators, only that of the MGC products that were extensively tested it responds best in the usual environmental conditions of anoxia treatment.)

The reliability of color change in Ageless-Eye® after reuse and prolonged air exposure was evaluated over a thirteen-month period. The indicators were exposed to forty-two air and low-oxygen cycles amounting to an air exposure of seventy-two days. All indicators continued to change pink and violet respectively, but the response to low-oxygen atmospheres became slower over the course of the test. Ageless-Eye® type K continued to turn pink within twenty-four hours, while type C required more than twelve days to change to pink at the end of the thirteen-month series of tests. Although these laboratory data indicate that Ageless-Eye® can be successfully reused repeatedly, the GCI does not recommend it. Certainly, with the cost of each Ageless-Eye® relatively low, there is no reason to take any risk or introduce uncertainty about the time a color change may take when an Ageless-Eye® is reused.

To test their performance after a long period of storage, both Ageless-Eye® types were kept in a low-oxygen atmosphere at about 45% RH for one year. The indicators retained their activity as shown by changing to violet within twenty minutes after they were exposed to air, with the exception of one tablet out of approximately fifty. The conclusion is that MGC's statement of guarantee of safe storage of Ageless-Eye® for six months is very conservative. However, to be sure that the activity of Ageless-Eye® that has been stored for a long time is not adversely affected by any unknown factors, a check is advisable, perhaps monthly after six months. This is easily performed by taking two samples, exposing them to air and timing the color change, then resealing them with Ageless oxygen absorber. If their color reverts to pink within a day, they are safe to be used as oxygen indicators in anoxia treatment.

When purchasing the indicator, the individual who is planning anoxia should require information about the storage conditions of the Ageless-Eye® types available. If the response of the seller is not satisfactory, any Ageless-Eye® bought from the supplier should be tested at the time of receipt, and perhaps every two months thereafter, by the procedure listed above. Of course, a record should be kept so that any increase in the time required for color change can be verified.

The supply of Ageless-Eye® has been limited in the United States and Europe, but small quantities are available through conservation supply companies.

It is not possible to estimate the freshness of the indicator as the packages do not have a date stamp. (Providing a date stamp has been suggested to the manufacturer.) The reaction and reversibility test should be performed on Ageless-Eye® samples to ensure their reactivity before use in anoxia treatments.

### Oxygen monitors and analyzers

Oxygen detectors are described below in three broad categories: less costly monitors that are adequate for safety purposes but are not sensitive enough for all oxygen analysis in anoxia containers, low-end scientific analyzers that are recommended for monitoring anoxia treatment enclosures, and high-end analyzers for research or measurement of minute oxygen change in anoxia work. It is advisable to purchase at least one low-end oxygen monitor (or a more expensive head-space analyzer) no matter what anoxia procedures are to be followed. The specific oxygen concentrations that a monitor can detect and display provide objective data about the adequacy of the seals in bags and confidence that the low levels of oxygen needed for anoxia are attained. A monitor can also be used to ensure that visual indicators such as Ageless-Eye® are performing correctly.

Simple oxygen monitors or indicators (Fig. 2.7) can be used for rough or approximate indications of the oxygen levels inside an anoxia enclosure, not generally for accuracy to 0.1% oxygen. However, their primary application is to ensure the safety of individuals in confined workspaces where oxygen deficiency is a possibility. Although nitrogen is not toxic, its displacement of the oxygen in a near-static atmosphere can decrease the oxygen concentration so much that impaired thinking and motor activity occurs—prior to more serious physiological events. Accidents resulting from human anoxia are impossible with a cylinder of nitrogen slowly flushing, say, a 200 L (7 $ft^3$) anoxia bag in a room with normally rapid forced-air circulation. However, where a large tent might be purged of oxygen by a nitrogen flow of 100 L per minute (see Chapter 4), safety monitors are essential to warn of a serious decrease in the oxygen content of the room or a portion of it.

Pocket-size monitors that can be carried and also mounted in the work area (see Fig. 2.7) usually measure the range of 0% to 25% oxygen, readable to 0.1%. An alarm in the monitor notifies the bearer or, if the monitor is mounted, the persons nearby when the oxygen concentration falls below a safe limit of 19.5% (OSHA's safety limit). If the monitor is to be used to measure the oxygen concentration in an anoxia enclosure it is important to realize that although the display might show 0.3% oxygen, the accuracy is not guaranteed to be better than within several tenths of a percent of that amount.[4] Thus only an approximation of the true oxygen concentration inside the anoxia chamber could be obtained, even though the instrument is perfectly adequate for safety purposes in the room outside it. All such monitors purchased for assuring safety should have a built-in temperature compensation and be capable of calibration in air, that is, at 20.9% oxygen. Their cost is about $525.

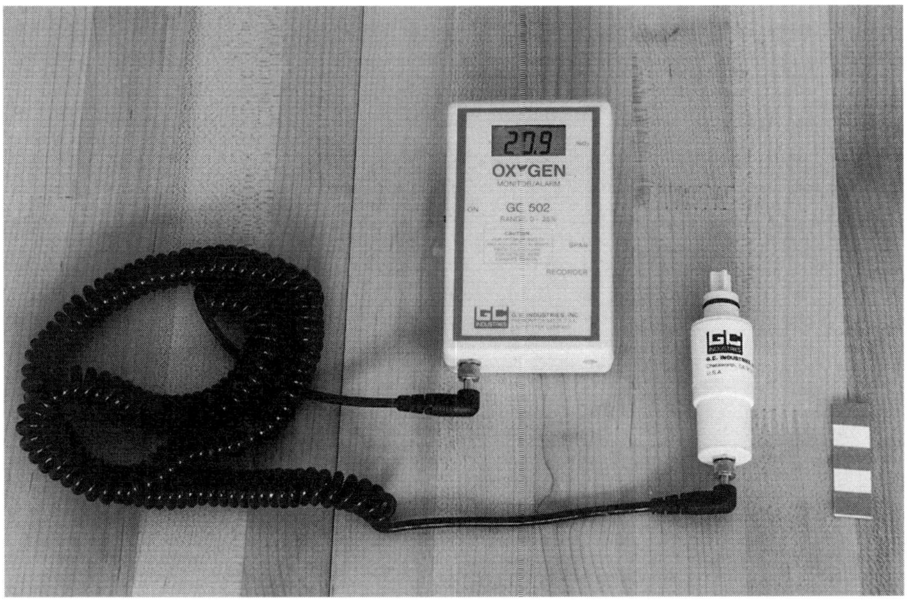


**Figure 2.7**

Oxygen monitor, with alarm, used for safety (scale = 5 cm)

During the testing of procedures and development of techniques for anoxia at the GCI, relatively inexpensive portable oxygen analyzers (Fig. 2.8) have been the instruments of choice for measuring oxygen in anoxic treatment containers, from medium-sized bags to large tents. Because they are sensitive and accurate, with a resolution and accuracy of 0.02% oxygen in a range from 0% to 1%,[5] they clearly show when optimal conditions for anoxia (of less than 0.3% of oxygen) are reached. Just as important, they are a sure indication if those conditions are maintained over the period of time required for anoxia of insects. Portable analyzers use the same cells as the more expensive trace analyzers described below but have the additional advantage of being powered either by rechargeable batteries or by connection to an external power source. If the analyzer is battery operated, it can simply be left inside the enclosure and any changes in the oxygen concentration directly observed on its display. (When an aluminized film is used for the enclosure, of course, a viewing window must be sealed in it.) An alternative procedure is to place the oxygen sensor of the analyzer inside the anoxia bag and pass wires from it through a hermetically sealed connector (or a hole that is skillfully sealed with an oil-based sealing putty that is never disturbed during the anoxia period) in the enclosure film to the analyzer (Fig. 2.9). Portable analyzers are available for $1,000; sensors are below $200.

There is a type of portable analyzer that is more convenient and versatile than the above-described units: a head-space oxygen analyzer. This instrument is commonly used in the food industry to perform spot checks to determine the oxygen content inside oxygen-free packages that contain snacks such as potato chips. The procedure is remarkably easy. Using a special syringe needle attached to tubing, an operator simply punctures a bag of chips through a sticky patch placed on it. A small pump in the analyzer draws a sample from the bag's atmosphere through the tubing to its sensor until the reading stabilizes. Thereupon, the oxygen content of the bag is digitally displayed. Although the GCI has tested head-space analyzers less extensively than other modes of oxygen determination in its years of work with anoxia, the convenience and

**Figure 2.8, below**
Portable oxygen analyzer (scale = 5 cm)

**Figure 2.9, below right**
Portable analyzer attached to anoxic treatment bag via hermetically sealed connector

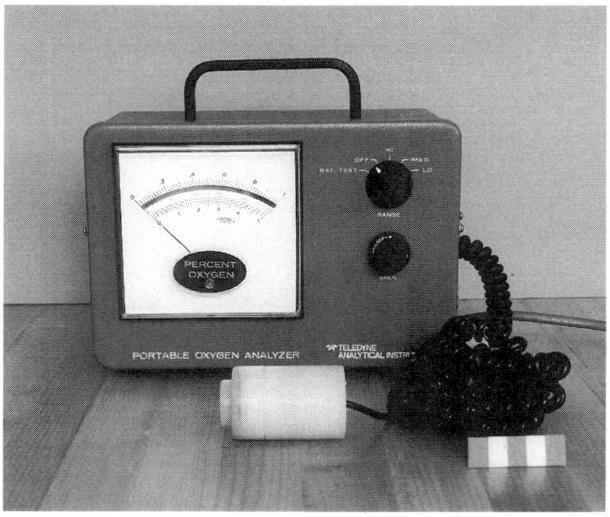


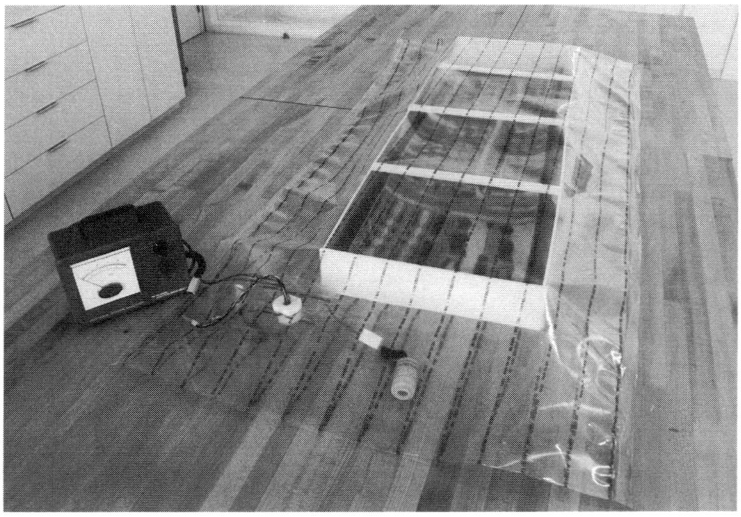

versatility of being able to quickly learn the oxygen contents of several (or many) anoxic bags commend this kind of monitor to conservators.

The least expensive head-space analyzer is a palm-sized, hand-held instrument. At least two of the several brands on the market have accuracy such that a 0.30% oxygen reading on their display is ± 0.03%. Therefore, when the oxygen level in an anoxia enclosure reaches 0.30%, one can be sure that this is the maximum oxygen that can be tolerated. Some brands are accurate only to ± 0.05%. With these, the danger point for oxygen level in anoxia should be taken as 0.25% because waiting until a reading of 0.30% might mean an actual 0 35% oxygen content. Costs of hand-held head-space analyzers are approximately $3,000.

Larger head-space analyzers, which are still as portable as a moderately heavy radio, offer even more advantages. At least one has a built-in printer (Fig. 2.10) so that oxygen concentrations in a bag's atmosphere are not only displayed but recorded (with time and date as well). All have RS 232 outputs for linkage to a computer. Prices of the portable units are in the $5,000 range. An additional advantage of one brand of portable head-space oxygen analyzer is its use either for manual sampling of the oxygen level in an enclosure, that is, samples taken at any chosen time by a user, or for automatic ("robotic") sampling at set intervals—from minutes to days. Although a number of sequential samples can be taken if a syringe is just left in its sticky septum in a bag, there is a much better technique. This involves attaching a septum holder through the enclosure wall and inserting the syringe needle through a silicone-rubber septum for the duration of the anoxia treatment. (As shown in Fig. 2.10, below, but with a gas chromatographic septum, rather than tubing, held tightly in place by the Swagelok nut.) A graph of the oxygen level throughout an anoxic treatment can then be made with the aid of the appropriate computer software.

Trace analyzers (Fig. 2.11) are both sensitive and accurate in determining very low levels of oxygen in the parts per million range up to 1%. They are required for research in anoxia in which precise measurements are essential. Two modes of operation are possible in anoxia investigations: (1) an oxygen sensor can be placed in the enclosure used for anoxia and its

**Figure 2.10, below**
Head-space analyzer with printer

**Figure 2.11, below right**
Trace oxygen analyzer (scale = 5 cm)

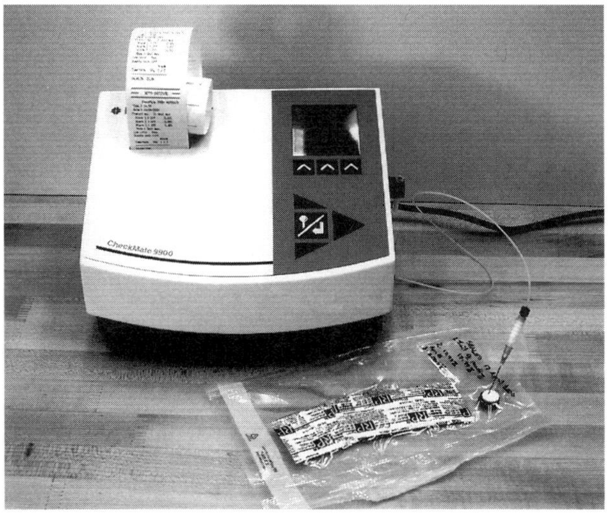

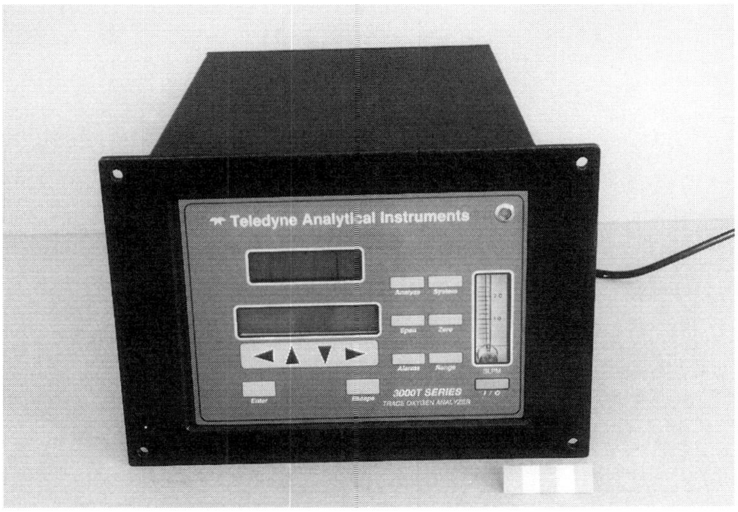

wires passed through the container walls via hermetically sealed connectors to the analyzer; or (2) a stainless steel diaphragm pump connected by tubing sealed to the container can draw a flow from its atmosphere through the trace analyzer sensor and back to the container, if it is less than 5 $m^3$, or vented to the room or outside, if it is larger. The first procedure is far superior to the second because it can result in a continuous record of the oxygen content in the enclosure and is passive; that is, it does not require an operator to perform the analysis at intervals.

In the portable analyzer, the remote sensor, or the trace analyzer, the older type of device responsible for quantitatively detecting oxygen is an electrolytic cell consisting of cathode, anode, and electrolyte. When oxygen in any gas entering the cell (by diffusion through a membrane) contacts the electrolyte solution, an electrochemical reaction ensues whose effect is proportional to the amount of oxygen present. This can be converted into a minute signal, amplified, and displayed in a digital or analog format. Because these electrolytic cells undergo depletion in the presence of air, even when stored on a shelf, they have a limited life, and they must be replaced regularly. In the past decade, more expensive zirconia cells have begun to dominate because of their nondepleting nature and long service life. The cost of sensor cells is between \$100 and \$200 and of a trace analyzer about \$3,500.

### Relative humidity monitors

Many types of RH monitors are available, ranging from inexpensive indicator strips to traditional hair hygrometers and including modern instruments based on the resistivity of bulk polymers and on capacitance, as well as research-level chilled-mirror apparatus (Fig. 2.12). The costs are from below \$75 to \$5,000, but there is little need for an instrument costing more than several hundred dollars. Most instruments respond well in the range of 20% to 80% RH; some are slow. The selection of a monitor

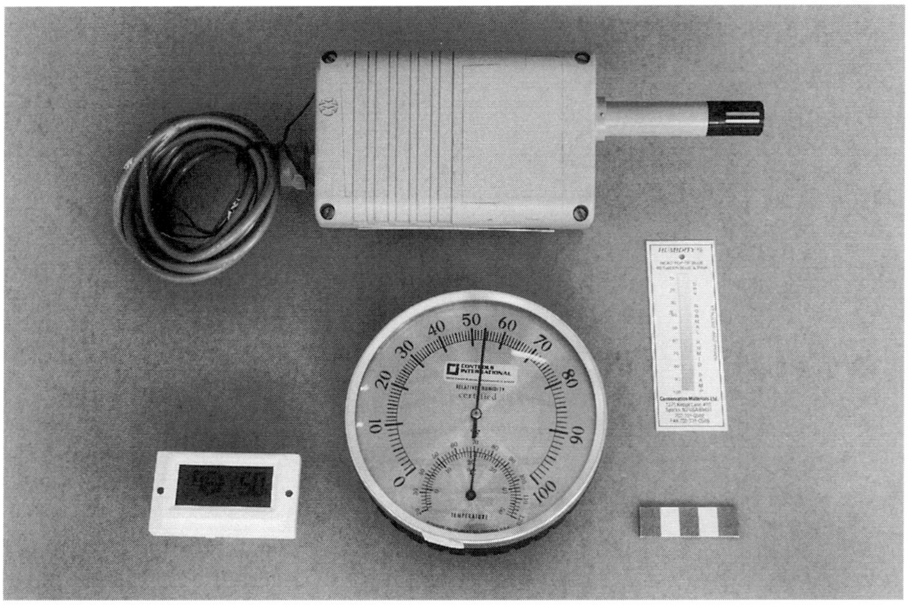


**Figure 2.12**
Capacitance sensors, digital or analog hygrometers, and indicator strips for measurement of relative humidity (scale = 5 cm)

type for use in anoxia bags should be based primarily on accuracy, convenience of use (which includes size), and cost. In addition, response time is important when nitrogen is being humidified in flowing through all sizes of anoxic enclosures (see Chapter 4). Yet clearly, quick response is unnecessary for weeks-long observation of the atmospheric relative humidity in small-scale anoxia.

Measurements in small sealed bags and pouches, where changes in relative humidity are minimal and usually occur slowly, can be done even with inexpensive indicator strips. However, modern hygrometers with unambiguous digital readout, available from any chemical supply company such as Cole-Parmer, cost less than $75 and are permanent. Many other more accurate units are less than $400. The less expensive instruments either cannot be calibrated or can be calibrated[6] at only one relative humidity, whereas the more costly yield more accurate results because of their calibration at both high and low relative humidity points.

It is primarily when nitrogen must be humidified to flush bags or tents that accurate and fast humidity monitors are required, first, in the calibration of the rate of flow versus relative humidity, and second, in the actual flushing process. Especially during the initial purging of tents when thousands of liters of humidified nitrogen are used, a drift or slow response in a hygrometer can result in an unwanted drastic change in the true relative humidity of the tent's atmosphere. This could produce RH shock in moisture-sensitive objects within.

Although bulk polymer sensors with digital readout cost less than $45, they take up to five minutes to produce 90% of their ultimate RH reading. In contrast, most monitors based on capacitance sensors generally provide similar readings in ten to thirty seconds for only $10 to $15 more. Even though battery operated, their function and display require so little power that usually no battery replacement is necessary for a year. One should be careful to purchase a monitor that either reads the RH continually (if it is to be placed inside the anoxia enclosure) or has a probe that can be installed in the bag (with proper sealing of the wires through its side) with RH readable on the external instrument. Of course, any monitor that is placed in a bag for a fourteen-day anoxic treatment must have the battery capacity for that time span.

Duct-mounting temperature and humidity sensors, whose sensing elements are mounted at the tip of 10 to 15 cm long tubes extending from the rest of the electronics, are used in heating, ventilation, and air-conditioning (HVAC) applications. These sensors are obtainable from HVAC repair or parts suppliers and can be effectively employed for monitoring relative humidity in bags and gas streams in anoxic treatment.

Indicator strips are the least expensive method of monitoring the relative humidity in an anoxia enclosure ($50 for a package of ten). Hair hygrometers cost from $50 to $700, and high-end units can provide an inked record of a three-week anoxia treatment, both temperature and relative humidity. Electronic sensors of all types are available from many sources such as those listed in Appendix IV.

### Temperature sensors and thermometers

Because temperature is one of the factors influencing insect mortality, monitoring is advisable. (Cautions in Chapter 1 about higher temperatures harming museum objects and temperatures lower than 25°C aiding insect survival cannot be overemphasized.) If the treatment is carried out in a temperature-controlled room, no monitoring inside the enclosure is necessary. Nevertheless, the process of noting or recording the temperature is so easy (and part of so many humidity instruments) that there is little reason not to do it.

More varieties of temperature sensors are available than of any other types of measuring devices or instruments: temperature indicator strips and paints, bimetal or dial thermometers, glass thermometers, thermocouples, and thermistors. All of the foregoing, except temperature indicator strips and paints, are accurate enough for anoxia. Glass thermometers are inadvisable in a workplace such as that for anoxic treatment because they break readily and release mercury.

Bimetal dial temperature indicators are rugged and cost less than $25. Thermocouples and thermistors are electronic sensors that can be connected to chart recorders or dataloggers for automatic monitoring. K-type thermocouples and resistive thermal devices (RTD) are the least expensive electronic thermometers ($40). In general, thermistors are more accurate than thermocouples, but they cost more.

### Leak detectors

Anoxia enclosures should be as gas-tight as possible. Therefore, testing to find leaks and eliminate them is essential. The greatest risk of leakage is along seams—including those for windows in aluminized film, access ports, and around connectors—or through caulked holes through which wires go to sensors. Generally, leak detection becomes easier when the enclosure is slightly pressurized using air or nitrogen.

**Caution!** Never exert more pressure than necessary to barely inflate a bag during anoxia. Portions of seams can be opened and major leak problems caused by excessive pressure.

The detection of leak points can be accomplished in several different ways. For rigid chambers, but not for flexible bags and tents, leak testing can be done ultrasonically. A battery-operated ultrasonic generator (35,000–40,000 Hz) is placed inside the container, the chamber's access port is clamped, and then the ultrasonic detector is passed over the chamber searching for leaks. (The access port is a good initial test site.) Equipment of this type runs from $500 to $1,000. A detection system based on visible fluorescence works similarly for either rigid chambers or flexible bags. A tracer gas is released into a bag by injection through the laminate film with a gas syringe, and after taping the injection spot, when an ultraviolet light is played over the surface, leaks show themselves as a fluorescent glow where the tracer gas escapes. (The variation for a rigid chamber is simply to squirt a bit of tracer gas before quickly shutting and tightening the door.) The cost for this fluorescent leak detector is between $500 and $800. The least expensive and most popular method is using a halogen leak detector and a tracer gas (Fig. 2.13) such as tetrafluoroethane (air

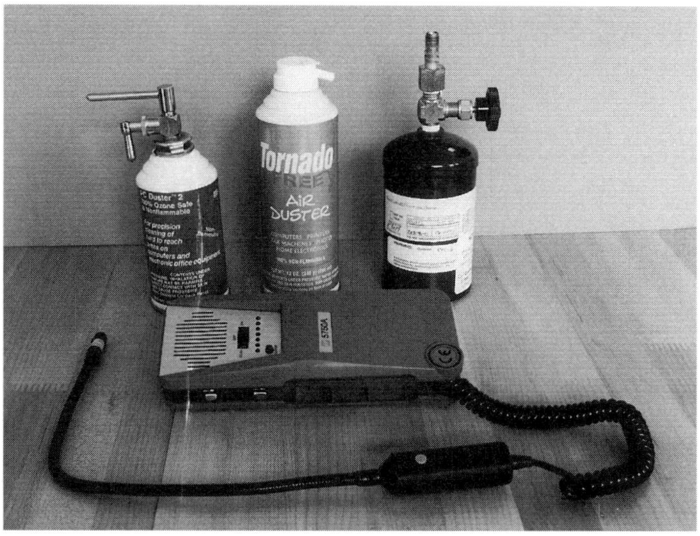


**Figure 2.13**
Leak detector and tracer gas (scale = 5 cm)

dusters, computer cleaners, or refrigerants). The tracer gas causes the thermal conductivity in the atmosphere to change at the point of leakage. Thus the sensor attached to the detector nozzle senses such a change and emits a characteristic alarm. This kind of detection equipment ranges from $200 to $500, depending on the type of tracer gas and the manufacturer. Tracer gas systems work well in all types of enclosures.

### Data collection during the anoxic process

Management of the data obtained from measuring the oxygen concentration, humidity, and temperature during anoxia treatments depends, of course, on the resources and the preference of the person doing the work. Everyone is aware of the usefulness of some kind of history of a procedure, both for checking later to see if conditions were not according to recommendations and for assisting a novice who at some time must perform a similar operation. In addition, such records are essential for formal reports to management or colleagues. However, the many tasks with which everyone is faced make it imperative that only the most important new jobs be added.

Recording all the data points obtainable in an anoxia process can fill too many notebook pages. Of course, a good written record must be kept of the actions, the materials used, the mishaps, and the successes in the process. This is essential. Oxygen, humidity, and temperature data can be hand-recorded at just a few intervals and still yield a useful guide for others to follow, or to interpret, or on which to base formal future reports. This is the minimum that should be done.

Monitoring is important and can be done inexpensively by using nonelectronic sensors. If funds permit, electronic oxygen, humidity, and temperature sensors should be placed inside all medium to large anoxia enclosures in a museum workplace. These can then be connected to dataloggers and the collected data automatically processed using simple spreadsheet computer programs. When purchased, most dataloggers are packaged with specialized software that will output data on a computer as curves over the entire treatment period. Taped in the anoxia notebook, this is an important time saver and is the best way to see precisely what happened to vital parameters during the treatment process.

Integrated sensor-datalogger suppliers include Omega Engineering, Inc., Cole-Parmer Instrument Co., and Mitchell Instrument Co. Conservation Support offers several types of inexpensive loggers. Other suppliers and manufacturers of the types of equipment described in this chapter are listed in Appendix IV.

Those involved in anoxia work are urged to consult the print or on-line catalogs of the companies listed in Appendix IV for current and precise information on specifications, including measuring range, resolution, accuracy, calibration capability, power supply requirements and options, datalogging capability, and costs.

## Notes

1. The transmission of gases by a polymer film can be decreased by stretching the film while it is heated. This orients the long polymer chains so that there can be more sites of molecular attraction among them than in their prior, more random state. Some of the more common films that are so treated are PET, PVDC, PP, and cross-linked PE. Nylon fibers are the classic original example of strengthening due to stretching.
2. See Note 1.
3. Multisorb Technologies, Inc., manufactures oxygen indicators in tablet form that change from a pale yellow to green in the presence of oxygen. These indicators are currently not sold separately but may be available from IMPAK. They have not been evaluated for museum use.
4. The guaranteed accuracy for monitors with a measuring range of 0% to 25% oxygen is ±2% of the full range, i.e., a difference of ±0.5% oxygen. Therefore, the monitor is warranted only to measure an actual oxygen concentration of 0.5% to ±0.5% accuracy—possibly the display could read 0.0% or 1.0% for this 0.5% oxygen level. This is not accurate enough for determining the actual oxygen concentration during anoxia treatments, even though such monitors often may be within 0.1% of the true concentration. However, they are accurate enough in the 19.5% range to rely on them for safety purposes.
5. Some instruments allow the operator to select from different measuring ranges, for example, 0–1%, 0–10%, or 0–25% oxygen concentration.
6. RH sensors can be calibrated by exposing them to a constant relative humidity for a prolonged period. Saturated solutions of different salts are used in airtight enclosures to provide a stable RH environment. Information on the relative humidity of various saturated salt solutions can be obtained from chemical handbooks. The salt should be only slightly moistened and placed in wide containers. Of course, the sensors themselves should be protected from contact with salt or solution to avoid contamination.

## References

**American Society for Testing Materials**
1999. ASTM D3985-95—Standard test method for oxygen gas transmission rate through plastic film and sheeting using a coulometric sensor. In *1999 Annual Book of ASTM Standards*, vol. 15.09, 489–94. West Conshohocken, Pa.: ASTM.

**Arney, J. S., A. J. Jacobs, and R. Newman**
1979. The influence of oxygen on the fading of organic colorants. *Journal of the American Institute for Conservation* 18:108–17.

**Ash, M., and I. Ash**
1992. *Handbook of Plastic Compounds, Elastomers, and Resins*. New York: VCH Publishers.

**Benning, C. J.**
1983. *Plastic Films for Packaging*. Lancaster, Pa.: Technomic Publishing Co.

**Briston, J., ed.**
1983. *Plastics in Packaging—Properties and Application, Institute of Packaging*. Melton Mowbray, Leicestershire: Whitefriars Press.

**Calmes, A.**
1985. Charters of Freedom of the United States. In *Showcases: Museum*, vol. 146, 99–101. Paris: UNESCO.

**Corish, P. J., ed.**
1992. *Concise Encyclopedia of Polymer Processing & Applications.* Oxford: Pergamon Press.

**Cryovac Division, W. R. Grace & Co**
1996. *Surlyn Product Data Sheet (TO50B-5).* Duncan, S.C.: W. R. Grace & Co.

**DuPont**
1994. *Preliminary Technical Information, Mylar® 50XMMC3.* Wilmington, Del.: DuPont Packaging and Industrial Polymers Division.

**Finlayson, K. M., ed.**
1989. *Plastic Film Technology—High Barrier Plastic Films for Packaging, Volume 1.* Lancaster, Pa.: Technomic Publishing Co.

**Gilberg, M.**
1989. Ageless Oxygen Scavenger. *Australian Institute for Conservation of Cultural Material National Newsletter* 32.

**Goodfellow**
2000. Product Information. *Online Catalog*: www.goodfellow.com.

**Green, L. R., and D. Thickett**
1995. Testing materials for use in the storage and display of antiquities: A revised methodology. *Studies in Conservation* 40:145–52

**Kenjo, T.**
1980. Studies on the long-term conservation of cultural properties: The effects of different concentrations of oxygen on pigments used for cultural properties. *Scientific Papers on Japanese Antiques and Art Crafts* (translated from Japanese) 25:103–7.

**Koestler, R., E. Parreira, E. Santoro, and P. Noble**
1993. Visual effects of selected biocides on easel painting materials. *Studies in Conservation* 38:265–73.

**Kuehn, H.**
1968. Watercolors: Reduced light-damage in a nitrogen atmosphere. In *Contributions to the London Conference on Museum Climatology 1967*, ed. G. Thomson, 79–88. London: International Institute for Conservation of Historic and Artistic Works.

**Ludlow Corporation**
1997. Data sheet for Filmpak 1193. Homer, La.: Ludlow Corporation.

**Maekawa, S., F. Preusser, and F. Lambert**
1992. An hermetically sealed display and storage case for sensitive organic objects in inert atmospheres. In *Biodeterioration of Cultural Property 2: Proceedings of the 2nd International Conference on Biodeterioration of Cultural Property*, ed. K. Toishi et al., 374–85. Yokohama: International Congress of Biodeterioration of Cultural Properties.

**MGC (Mitsubishi Gas Chemical Co.)**
1994. *Ageless® Oxygen Absorber: A New Age in Food Preservation.* Tokyo: Mitsubishi Gas Chemical Co.

1997. *RP System™ For Cultural Artifacts: Usage Manual.* Tokyo: Mitsubishi Gas Chemical Co.

**Paine, F. A., ed.**
1991. *The Packaging User's Handbook.* Glasgow: Blackie & Son.

**Plastics Design Library**
1995. *Permeability and Other Film Properties.* New York: Plastics Design Library.

**Sweeting, O. J., ed.**
1971a. *The Science and Technology of Polymer Films, Volume 1*. New York: Wiley.

1971b. *The Science and Technology of Polymer Films, Volume 2*. New York: Wiley.

# Chapter 3

## Small-Scale Anoxia Using an Oxygen Absorber

For small-scale anoxia, oxygen is removed from the atmosphere in the bag by an oxygen absorber that maintains the oxygen level below 0.3% for the fourteen-day treatment period. This procedure in small to medium-sized bags is especially useful when a limited number of objects are to be treated. (It is more expensive than flushing air from anoxia bags of about 300 L and above with humidified nitrogen if many anoxia treatments are to be done.) Step-by-step instructions for fabricating plastic bags and carrying out anoxia using oxygen absorber are provided in Chapter 5.

This chapter begins with an introduction to the chemical nature of the most common type of oxygen absorber, continues with a survey of the other oxygen absorbers that are commercially available, and reports the findings of GCI investigations into their properties. We then show how to calculate the amount of an oxygen absorber needed for an anoxia treatment, describe some facets of anoxia in plastic bags, and discuss the problem of controlling moisture release by oxygen absorbers during anoxia and propose a solution.

### The Chemical Process of Oxygen Absorption

Commercial oxygen absorbers are composed of substances that react with oxygen and thus decrease its concentration toward 0.001% in totally sealed containers. They are packaged in small gas-permeable packets (sachets). Most of these absorbers use specially processed moist iron oxide powder as the active component (Plate 3). The oxygen absorption process can be thought of as the complete corrosion of iron, because the product is moist iron rust, or hydrated iron oxide.

Recently, the RP-System™, a different type of oxygen absorber, has been introduced to the market. It contains unsaturated organic compounds that absorb corrosive gases as well as oxygen. (One type also absorbs some water.) Unfortunately, the active compounds are proprietary and information on their reaction with oxygen is not available (MGC 1997a).

## Properties of Oxygen Absorbers: Absorption Capacity, Reaction Rate, and Moisture Release

Oxygen absorbers made from processed iron powder are nontoxic, and because they do not release harmful gases, they are safe to use with objects made of all types of organic and inorganic materials. However, the oxidation of iron in this powdered state is a rapid and notably exothermic reaction. Therefore, the oxygen absorber packets can become hot, especially if they are unwisely exposed to air for many minutes when stacked or piled together. While the generated heat of a moderate number of packets spread out in an anoxic enclosure does not cause a significant increase in the environmental temperature and a consequent drop in relative humidity, absorber packets piled directly on an object could cause damage to it.

The oxygen absorption process is moderately temperature dependent and thus will be more rapid at higher temperatures. (This underscores the importance of avoiding piles of packets.) As an aid to any solid-gas reaction, it is wise to expose the maximum surface area of the iron-filled packets to the oxygen in the atmosphere of the anoxia enclosure. An effective arrangement is to place them edge up and separated by mesh spacers in a plastic box. Individual packets can be stapled into a short or long chain to make it easier to slide them into the bag and to avoid having a number of packets stacked together in the bag.

Several companies are currently producing oxygen absorber, for example, Ageless® by Mitsubishi Gas Chemical Company in Japan, FreshPax™ by Multisorb Technologies in the United States, and Atco™ by Standa Industrie in France (Plate 4). Absorbers are sold in sizes, noted on the individual packets, that state the amount of oxygen measured in cubic centimeters that they are capable of absorbing, as in Ageless® Z-1000, FreshPax™ D-1000, Atco™ LH 1000. Each of these is guaranteed by the maker to absorb 1000 cc (mL) of oxygen.

The desire to ensure that the products will perform well, even if mishandled by the consumer, has led Mitsubishi to put three times the nominal amount in their packets (Kondoh 1995) and Standa, about 50% more. The advantage of this is that even if Ageless® Z-1000 is exposed to air for as long as several minutes before it is placed in an anoxia chamber, or its storage conditions are somewhat less than perfect, it will still surely absorb more than its rated capacity of a liter of oxygen. (The user can work at a measured and deliberate pace, without having to be concerned about a few minutes' delay, rather than hurriedly try to protect the oxygen absorber from any but a few seconds of contact with air.) The disadvantage is that a knowledgeable individual may calculate precisely on this overcapacity packaging but the company might choose not to continue it. Further, the GCI's quantitative scientific data concerning the rate of reaction for Ageless® Z is based on a 1988 batch that was assumed to contain the nominal packet capacity (Lambert, Daniel, and Preusser 1992). Thus all calculations using those data are conservative by a factor of at least two; they can be changed to reflect such an apparent increased rate of reaction.

Oxygen absorbers are available in sealed packages that contain a number of individual sachets with a total capacity of absorbing between 10,000 and 50,000 cc of oxygen (e.g., a few Ageless® Z-1000 or FreshPax™ D-1000, or many Z-300 or LH 100, etc.) (Fig. 3.1). They can be stored at normal temperatures. Obviously, care must be taken that the master package is well sealed and not accidentally punctured or torn. An oxygen indicator tablet inside the package verifies the oxygen-free condition of the packets when it is pink in color rather than violet. Although MGC states that exposure to air of up to four hours is acceptable for Ageless® Z, it recommends that the absorber be used as quickly as possible (MGC 1994). Experience has shown that the original package can be easily and effectively resealed after removal of some of the absorber packets for use in anoxia.

Water is essential to speed the reaction of iron powder with oxygen. Therefore, the various brands of this type of oxygen absorber include water that is loosely held by a diatomaceous earth or zeolite or amorphous silica that acts as a moisture binder. In a 1000 cc packet, Ageless® Z has about 3.3 g of water, Atco™ LH 1.6 g. In FreshPax™ D-200, there are 2.9 g of water. Only part of the water in the binders in these sachets is used for the reaction with oxygen, and therefore, over time, some water is released to the environment in vapor form. This leads to an increase in the relative humidity of the atmosphere in an anoxia enclosure. To counter this and maintain a level of about 55% RH, a bed of silica gel in a plastic tray should be placed in the anoxia container. (Calculations for the amount of silica gel to be used are shown below.)

The manufacturers of oxygen absorbers sell only lots of several hundred packets with a capacity of 1000 cc, or an equivalent oxygen capacity of smaller and larger packets, for example, twice as many of 500 cc capacity, and so on. Thus, for the average conservator, a more practical source is a conservation materials supplier. Appendix IV lists several companies that sell various types of materials for anoxia; not all regularly stock oxygen absorbers.

Small amounts of oxygen absorber can be discarded through ordinary disposal methods, but the user must be warned of the hazard of putting several sachets, which could become hot, in a wastebasket containing flammable trash. It is a far better practice to gather any used packets and put them, separated as much as possible, in a loosely tied polyethylene plastic bag to allow them to react slowly with air. MGC strongly advises that absorber sachets must not be placed near combustibles and that sachets should not be torn open in an attempt to spread out the contents. Undoubtedly other manufacturers issue similar important warnings.

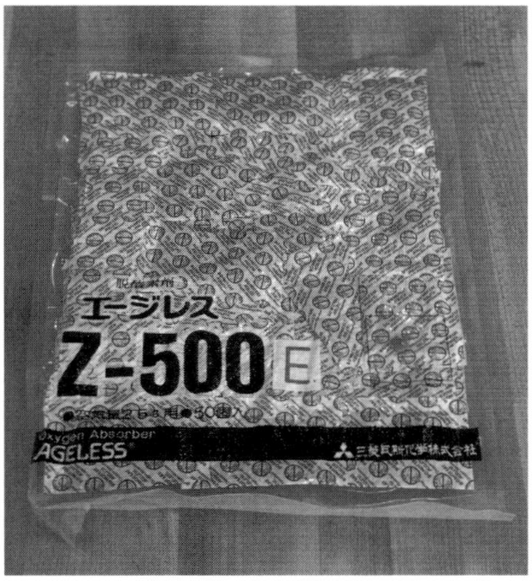


**Figure 3.1**
Oxygen absorber packets are vacuum sealed in plastic bags that have low oxygen permeability to prevent premature reaction with oxygen (scale = 5 cm).

## Ageless® oxygen absorbers

Until now, Ageless® has been the most widely used oxygen absorber in museums. Most of the GCI's scientific work on oxygen absorbers since 1988 has employed the Mitsubishi brand. This is not to imply that products from other manufacturers are not equal to Ageless® in performance or desirability. Where differences are known, they are discussed below in the sections on other brands of oxygen absorbers. Ageless® is described only as a typical iron-based product.

MGC makes several different types of Ageless® so as to match the water activity number (equilibrium humidity) of the packaged food products in which their oxygen absorber is used in very large quantities worldwide (Table 3.1). As seen in Table 3.1, Ageless® Z, specified to be used with products with a water activity number of 0.85 or less, would be best in anoxia treatments in which the desirable equilibrium humidity is approximately 55% (i.e., a water activity number of 0.55) or up to 80%.

The performance of all absorbers will degrade if used in atmospheres with RH levels much below 50%. This is especially true after months of such exposure, because the absorber will lose to the environment part of the moisture required for the oxidation reaction. A yearlong experimental test showed that Ageless® Z was capable of absorbing oxygen at a constant rate for about four months when exposed to 33% RH. Its reaction rate dropped to one-seventh of the original rate by the end of one year under this humidity condition

| Ageless® Type | Features | Application/ RH Conditions | Deoxygenation Time at Room Temperature | Max. Exposure to Air |
|---|---|---|---|---|
| Z | Suitable for dry products, can be used with desiccants | 85% or less | 1–4 days | 4 hours |
| ZPT | Oil-resistant packet material | 85% or less | 1–4 days | 4 hours |
| FX | Suitable for moist foods, stable in open air | 85% or more | 0.5–1 day | 12 hours |
| S | Fast working not to be used with desiccants | 65% or more | 0.5–2 days | 1 hour |
| SS | Low temperature application | 65% or more | 0.5–2 days | 1 hour |
| E | Absorbs oxygen and carbon dioxide | | Differs with size | 4 hours |

**Table 3.1**
Ageless® types, properties, and application specifications

(Lambert, Daniel, and Preusser 1992). These results indicate that an oxygen absorber can effectively be used in anoxic treatments at about 30% RH because of the short time for treatment, two to three weeks. However, RH levels that are below 30% in an anoxic procedure might cause a much more rapid degradation of the absorber's performance. Therefore, its use cannot be recommended for anoxia in very dry atmospheres without further experimental data.

Manufacturers' statements that their oxygen absorbers are capable of reducing the oxygen concentration in enclosures to 0.01% or less is easily achievable in practice. The rate of reaction of Ageless® with a fixed volume of oxygen in an anoxic chamber of any sort, as well as the ultimate minimal oxygen level attainable, depends on three factors: the oxygen-absorbing capacity of the packets in the enclosure, the volume of the enclosure, and the leak rate of oxygen from the outside into it. Users will easily achieve and maintain the desired anoxic environment of less than 0.3% oxygen in a chamber if the proper tools and materials are used. This is because the enclosure will have a low oxygen leak rate when it is well made of oxygen-barrier film and the appropriate number of oxygen-absorbing packets per the manufacturer's labeled capacity is placed in it. In a 20 L bag containing air (and thus about 4.1 L of oxygen), five packets of Ageless® Z-1000 resulted in an oxygen level of 0.16 in less than a day and 0.04 in less than two days. This is overkill. As sold now, Ageless® with a nominal capacity of 4000 cc is adequate to remove the oxygen from this 20 L bag to less than a 0.3% level and cope with an oxygen leakage of 10 to 15 cc per day for the fourteen-day period of anoxia. This is calculated on an actual content of oxygen absorber of only 10% above the labeled quantity. (As stated before, current manufacturing practice is to include much more than this.) These results were typical of more than one hundred experiments at the GCI with well-made anoxic containers of all kinds.

To repeat: Although the many results just described involved the Ageless® brand of oxygen absorber, the goal of the experimental work was to find the general behavior of this kind of packaged oxygen absorber when reacting with oxygen. There is no implication that the products of other manufacturers are not equally useful. Some of these products are described below.

### Atco™ oxygen absorbers

Standa offers two types of oxygen absorbers, Atco™ LH and Atco™ HV. Both can be used at low and high relative humidities (Standa 1994). Atco™ HV is the faster-reacting oxygen absorber, and the manufacturer states that 0.1% oxygen concentration can be reached within twelve hours with it. Atco™ LH requires fifteen hours (Standa 2000). Limited tests at the GCI, under conditions comparable to those used with Ageless®, showed that Atco™ LH absorbers required forty hours to reach oxygen concentrations below 0.1%. These tests also indicated that the actual capacity of Atco™ LH exceeded the indicated capacity by about 50%; that is, Atco™ LH 1000 was capable of absorbing 1500 cc of oxygen.

### FreshPax™ oxygen absorbers

Multisorb designed four types of absorbers (Table 3.2). FreshPax™ M and B are intended to be used in atmospheres with more than 65% RH and therefore have limited applicability for museum objects (Multisorb 1996a, 1996b). FreshPax™ R is specified for use with refrigerated products or where a rapid deoxygenation is required and may be useful for specific museum applications. It is the FreshPax™ D, designed for products with a water activity number below 0.7, that would be suitable for most anoxic treatments at 50% to 70% RH. Packets of FreshPax™ D are claimed to have a somewhat greater capacity for absorbing oxygen than the nominal amount printed on the packets. Distributors of FreshPax™ in the United States, IMPAK and Royco, offer bags and cases that contain the number of packets matched to museum users' needs (rather than massive food packaging requirements) and are very reasonably priced.

### RP-Systems™ oxygen absorbers

In the mid-1990s the RP-Systems™ RP-A and RP-K, specifically designed for use with metals and electronic parts, were introduced by MGC. More recently, this product has been modified for applications with cultural artifacts. The RP-System™ differs from the iron-based oxygen absorbers that have been discussed in having unsaturated organic compounds as the active component. The advantage of these organic substances is that they do not generate a great deal of heat during the oxidation reaction. In addition, one type absorbs moisture rather than releases it. A disadvantage is their cost—up to ten times that of iron-based oxygen absorbers.

RP-A (MGC 1997c) is unsuitable for anoxic treatment of insects in museums because it is designed to absorb moisture and decrease the humidity in any enclosure in which it is placed. However, this property, in addition to its ability to absorb corrosive gases, could be very advantageous in maintaining a low-moisture, low-oxygen atmosphere for a valued metal object.

**Table 3.2**

FreshPax™ types, properties, and application specifications

| FreshPax™ Type | Features | Application/ RH Conditions | Deoxygenation Time at Room Temperature | Max. Exposure to Air |
|---|---|---|---|---|
| B | Suitable for moist or semi-moist products | 70% or more | 0.5–2 days | 12 hours |
| D | Suitable for dehydrated or dry products | 70% or less | 0.5–4 days | 2 hours |
| R | Suitable for refrigerated products, fast working | all | 4 hours–1day | 1 hour |
| M | For moist or semimoist products, can be used with carbon dioxide flush | 65% or more | 0.5–2 days | 12 hours |

RP-K (MGC 1997d) oxygen absorber is claimed to be capable of decreasing the oxygen concentration in a sealed enclosure from 20.9% to less than 0.1% within two days (ostensibly in an enclosure whose volume equals the nominal capacity of the absorber). RP-K also absorbs corrosive gases and decreases the level of sulfur dioxide, hydrogen chloride, and ammonia to less than 1 ppm in several hours (MGC 1997b). The manufacturer states that moisture will be neither absorbed nor released by the RP-K during reaction with oxygen, and therefore the relative humidity inside an enclosure will not be affected (MGC 1997b). The RP-System™ is sold in packets whose different oxygen-absorbing capacities are denoted by the volume of air in the package or bag in which they should be placed, not by the amount of oxygen in the bag.

***Warning***: The user should be aware that this is a significantly different rating system from the amount of oxygen absorbed by a packet of Ageless®. One packet of RP-5K is recommended for enclosures with a volume of 500 cc of *air*; the RP-5K can absorb only about 100 cc of oxygen. An Ageless® Z-100 packet would be used for the same type of 500 cc bag.

## Amount of Oxygen Absorber to Use for Anoxia in Bags

How can one estimate the number of oxygen absorber packets to put in an anoxia enclosure? If one uses too many, some of the relatively expensive packets are wasted. If one uses too few, the oxygen content may not be reduced to less than 0.3% or the oxygen concentration may not be maintained at that level for two weeks.

There are four variables that determine the answer. Fortunately, only two rather obvious factors are required for calculation in insect anoxia: the amount of oxygen inside the bag containing objects to be treated and the oxygen-absorbing capacity of the individual packets that will be used. A sample calculation is given in the next section.

The two additional variables that determine how much oxygen absorber is necessary are the leak rate of the anoxia enclosure and the rate of reaction of the oxygen absorber with oxygen. Neither variable need concern those doing anoxia work *if* the treatment enclosure is well sealed. Research over a number of years at the GCI has proved that bags of many sizes can be fabricated with leak rates only a little more than the OTR value of the laminated film. (See Chapter 4 and Tables 4.1, 4.2, p. 77) A calculation that shows why no additional oxygen absorber is needed to compensate for leakage through the film in a two-week period is provided below. (If the user wants to measure the leak rate of the anoxia bag, the procedure for determining leak rates can be found in Appendix II.)

There still remains a major variable, the rate of the reaction of an oxygen absorber with oxygen. Will the absorber reduce the normal 20.9% oxygen in the air of an anoxia enclosure to a 0.3% oxygen level in a reasonable time? To 0.1%? Can it react with oxygen fast enough at 0.1% to cope with the amount of oxygen that leaks through the film and still maintain the 0.1% oxygen concentration? The answers to these questions are yes, based on the hundreds of practical experiments at the GCI and on the determination of the chemical rate constant, described in Appendix II. It is the latter constant that also answers other questions about the amount of oxygen absorber needed.

### Calculation of the amount of oxygen absorber

1. Sufficient oxygen absorber must be put in an anoxic bag to react with the oxygen in it. How much oxygen absorber to use in insect eradication depends on the volume of the anoxia enclosure and its gas-tightness. The precise volume of the usual rectangular bag, flat on the table or floor, is difficult to calculate. However, the volume can be roughly (and very adequately) estimated. Multiply the length and width of the bag by the height resulting from draping polyethylene (or the oxygen-barrier film) over dummies of the object(s) to be treated as described in Protocols A and B. Then an equation that includes the percentage of oxygen in air (20.9%, or approximately 21%) is

$$\text{Amt. of oxygen absorber needed (in units absorbing 1000 cc of oxygen)} = \text{Length (m)} \times \text{width (m)} \times \text{height (m)} \times 210$$

(If the measurements of the bag are in **centimeters** instead of meters, the conversion factor must be **0.00021** instead of 210 in the above equation.)

Example: Start with a bag flat on the floor that is a rectangle of film with a length of 1 m and a width of 0.5 m. If the height with a dummy inserted in it is 0.2 m, then $1 \text{ m} \times 0.5 \text{ m} \times 0.2 \text{ m} \times 210 = 21$. Thus 21 packets of Ageless® oxygen absorber of a capacity that would each react with 1000 cc (or 11 of 2000 cc capacity) would be adequate to absorb the oxygen in that bag. This amount also would maintain the oxygen level near 0.1% for the fourteen days of anoxia treatment *if* the bag was well sealed.

The assumption that the bag is actually a rectangular box (with a uniform height equal to the height of an object located only in its center) results in a calculated volume considerably greater than actual volume in a relaxed bag. However, this is usefully conservative in the sense that it leads to overestimation of the amount of oxygen in the bag. Thus the amount of oxygen absorber called for will also be overestimated. Thereby both the time to achieve anoxic conditions will be shortened and an ample supply of unused oxygen absorber will remain to cope with normal leakage through the sides of the bag even when a low oxygen level is reached.

The first of the following two sections does not enter in the necessary calculations of oxygen absorber *if* the anoxia bag has been properly sealed. The second section also is only for information rather than the user's action, because the rate of Ageless® reaction with oxygen has been determined at the GCI.

2. Enough oxygen absorber must be placed in the plastic bag to remove any oxygen that leaks into it during the two weeks of anoxia treatment. Some oxygen from the outside air is transmitted through the film each day. This quantity is easy to calculate from the OTR. As stated in the

previous section, the calculation is supported by many experimental results at the GCI. Admittedly, if the film or its sealing to form a bag is faulty, some oxygen (from the air) will leak in through pinholes and bad seams; this quantity can only be found (and compensated for by additional oxygen absorber) if the leak rate of the bag is experimentally determined (see Appendix II). It is far better to take extra time and care in making a well-sealed bag.

Transmission of oxygen through the plastic film is equal to the film's total area multiplied by the OTR (per square meter per day) and by the days of the anoxia treatment. As an example, assume that the 0.5 m × 1 m × 0.2 m bag described above is made of Aclar, a CTFE laminate that has only fair oxygen-barrier qualities (OTR of 10.5 cc/m$^2$/day in air; see Chapter 2 and Table 2.3 (p. 22) for OTR in oxygen that is 4.8 times that in air). A superior water barrier, Aclar is sold under the brand names Filmpak 1177 and FR-7750. For a fourteen-day period of anoxia, the oxygen leakage in cc through the two surfaces, top and bottom, would be two times 0.5 m × 1 m × 10.5 cc/m$^2$/day × 14 days = 147 cc. *This is a trivial amount, easily removed by the oxygen absorber capacity, 21,000 cc (1000 cc/packet × 21 packets), provided by the conservatively large estimate of the bag volume.*

3. Sufficient oxygen absorber must be present when the oxygen concentration in the bag drops below 0.3% to react rapidly enough to maintain that low level. It has not been widely discussed that the amount of oxygen with which an oxygen absorber can react per day—its rate of reaction with oxygen, not its capacity—is drastically reduced at low concentrations of oxygen, say, below 0.3%. As is shown in detail in Appendix I (and as is intuitively obvious), to maintain a constant oxygen level, the rate of reaction of oxygen absorber at any oxygen concentration must equal the rate at which oxygen leaks into the anoxia bag.

A good estimate of the amount of absorber needed to maintain 0.1% oxygen concentration (from Appendix II) is less than one-tenth of the capacity of oxygen absorber required to react with all the oxygen originally present in the bag. MGC's practice is to put more oxygen absorber in each packet than its rated capacity. Thus there is more than an adequate reserve for reaction even at 0.1% oxygen with the calculated oxygen absorber for removal of oxygen from the air in a bag. Of course, this assumes a well-sealed bag with no more than a few hundred cc of oxygen leaking in over the fourteen-day treatment period.

## Conclusion

The oxygen in the air of a well-sealed oxygen-barrier bag of 100 L could be reduced to 0.1% and maintained at that level for fourteen days by insertion of twenty-one packets of Ageless® Z-1000 in the bag. Calculations of the type illustrated above are applicable to plastic bags of other sizes and to other anoxia enclosures but not to other brands of oxygen absorbers without experimental data.

The equation discussed in Appendix II gives the time required for a given quantity of Ageless® to react with oxygen. It deals with a 100 L bag that would, according to the calculations a few paragraphs above, require twenty-one packets of Ageless® Z-1000. The oxygen level will be reduced to 0.3% in about thirty-eight hours and to 0.1% in two days. This prediction is conservative. Low levels are attained more rapidly because, as stated above, since the time the rate of reaction was determined, more than the nominal amount of oxygen absorber has been placed in the packets by the manufacturers.

### Procedure for Anoxia in Plastic Bags Using Oxygen Absorbers

General information about this procedure has been presented in Chapter 2. Also, a number of specific suggestions about bag construction have been given in the discussion of heat-sealing equipment in Chapter 2. Step-by-step instructions for anoxia using an oxygen absorber in small to moderate-sized plastic bags, including fabrication of the bags, are provided in Chapter 5. Therefore, our goal in this section is to present a few hints and to emphasize some cautions about bag fabrication and use, before taking up the problem of moisture release by oxygen absorbers and the solution to that problem.

The plastic laminated films that meet the criteria for anoxia (Nakamura and Hoshino 1983; Standa 1994) are described in Chapter 2, and the quantitative reason for not using polyethylene or similar films is discussed in Appendix II. High-quality premade pouches and small bags are available from some conservation supply houses. Many kinds of smaller objects can be simply placed in these bags together with the oxygen absorber, oxygen indicator, and moisture buffering material. The special clamps that are offered by one company succeed well in firmly closing the pouches for which they were designed and thus avoid the problems of heat sealing. However, these clamps are not effective with thicker or thinner bags and may lead to leaks.

Fragile objects obviously require special care. It is true that bags can be carefully designed and made to fit around an artifact, but it is always advisable to fully protect the object from damage by the plastic film. The object can be wrapped in tissue so that the film will not touch it, but it must be emphasized that the bag volume will decrease by one-fifth when the oxygen is removed by the absorber. A simple way to reduce any pressure that might be exerted on objects by the bag as it shrinks in volume is to inflate it somewhat with nitrogen. (Excessive inflation leading to great pressure inside the bag could weaken seams.) Alternatively, objects can be better protected by surrounding them with a lightweight frame made of wood or cardboard (Fig. 3.2) or heavy wire bent to a form. For very large objects, a protective framework of easily joined polyethylene pipe can be built in an hour or less.

To ensure that the number of oxygen absorber packets is adequate or to speed the process of attaining 0.3% oxygen, individuals have slightly evacuated a flexible bag with a vacuum cleaner or a vacuum pump. This is dangerous. With the additional shrinkage of the bag's volume as oxygen is absorbed, the bag could be drawn so tightly that it almost shrink wraps objects in it. (In addition, any sharp points of objects could puncture a tightened bag.)

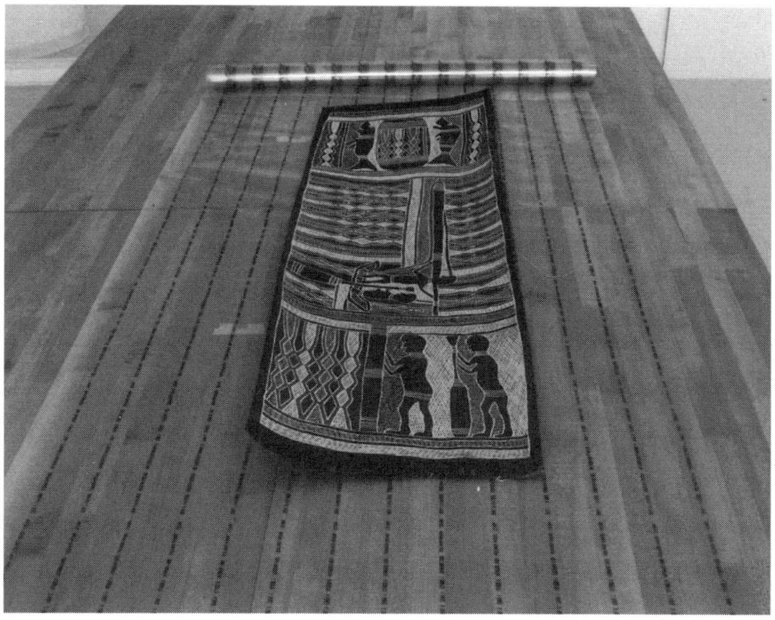

Purging with nitrogen gas is a preferred alternative to evacuation. Passing an amount of nitrogen equivalent to the volume of an anoxia enclosure through it will displace half of the oxygen. Nitrogen flushing requires only a modest amount of equipment, a cylinder of nitrogen, a gauge, and valves—unless the objects involved cannot tolerate a drop in the relative humidity of their environment even for two hours. In the latter case, humidification apparatus such as is described in Chapter 4 (and in more detail in Chapter 5) should be used to ensure the same relative humidity in the nitrogen that is passed into the bag as that which is best for the objects.

**Figure 3.2**
A frame protects fragile objects from being damaged by the plastic film.

## Control of Moisture Release by an Oxygen Absorber

Because the packets of oxygen absorbers contain more water than needed for their reaction with oxygen, moisture moves from them to the atmosphere of an enclosure over the course of anoxic treatment of objects therein. This property of Ageless® Z has been extensively investigated at the GCI. It is probable that other iron-based oxygen absorbers would behave similarly. When reacting with oxygen at 50% RH, a Z-1000 packet released 0.4 g of water. (Thus an Ageless® Z-500 will release 0.2 g, and other sizes do so proportionately.)

As is true of several other facets of anoxia treatment, the release of water from the oxygen absorber has good or bad consequences depending on the situation. As we have stated, flushing nitrogen through an anoxia enclosure immediately after insertion of the oxygen absorber packets is useful. This decreases the oxygen content of the atmosphere in the bag and thereby ensures that the amount of oxygen absorber calculated will be more than that required to achieve anoxic conditions. However, unless the nitrogen is humidified—a technique that is not difficult but requires some initial cost and appreciable labor—its flushing of the air from the enclosure may decrease the enclosure's relative humidity from an object's preferred 40% to 50% nearly to zero.

Of course, many types of objects are self-buffering against harm from a brief change in relative humidity. An additional aid is the relatively rapid release of water by Ageless®. Within three or four hours enough water is lost to the atmosphere by the packets—even in a large bag—to return the anoxic environment to 50% RH. (Only about 1.1 g of water is needed to bring a 100 L bag from 0% RH to 50% RH. The one-tenth or less of this amount of water that would be required for small bags means that 50% RH would be attained much faster in them.)

However, after this time moisture continues to come from the packets of oxygen absorber until an equilibrium relative humidity of

72% is reached. Over the two weeks of anoxia treatment, this level may be too high for certain museum objects and undesirable for others. Accordingly, a procedure for countering water emission from the oxygen absorber must be put in place to keep the relative humidity at a preferred level for two weeks.

Research indicates that silica gel, conditioned to the desired relative humidity, absorbs excess water rapidly enough to do an excellent job of coping with water that comes from oxygen absorber packets (Fig. 3.3). Figure 3.4 presents the results of a series of experiments with empty 16 L plastic bags whose air contents were originally at 50% RH. The top line is the RH change after insertion of four packets of Ageless® Z-1000; the lower lines correspond to other tests with the same amount of Ageless® but increasing amounts of silica gel in a plastic pan. Without silica gel, the relative humidity rises above 80% within an hour and gradually equilibrates at 72% in about forty-eight hours. However, with 160 g to 400 g of silica gel that had been conditioned to 50% RH (by lengthy exposure of dry silica gel to a 50% RH atmosphere), there was no change in the relative humidity. A more reasonable approach than attempting to hold the relative humidity precisely at the original value by means of such large amounts of conditioned silica gel is to use a moderate quantity that would keep it from rising so high. The conclusion is that 20 g of dry bead-type silica gel, equal to 25.4 g of silica gel conditioned at 50% RH (Fig. 3.5), is required for each Ageless® Z-1000 packet in order to limit the increase in relative humidity to 5%,

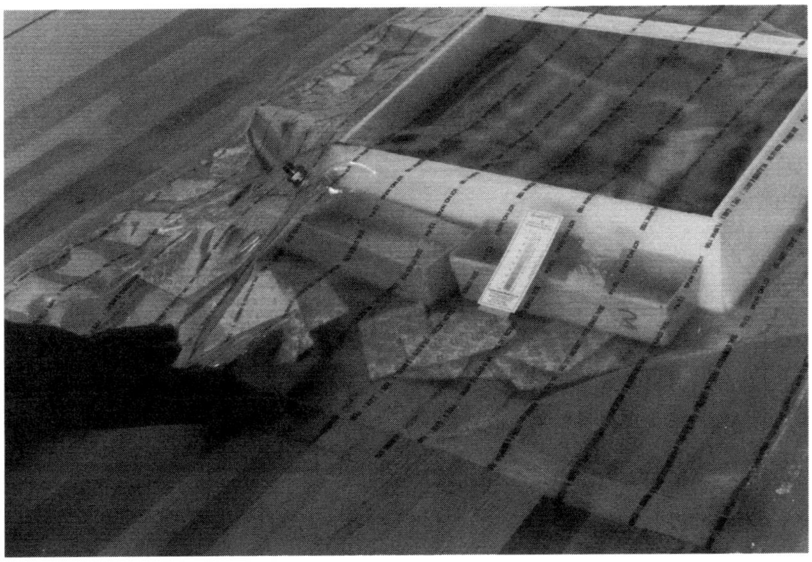

**Figure 3.3**
Oxygen absorber packets and humidity buffering material (RH-adjusted silica gel) are placed in the bag after most of the heat seals have been completed.

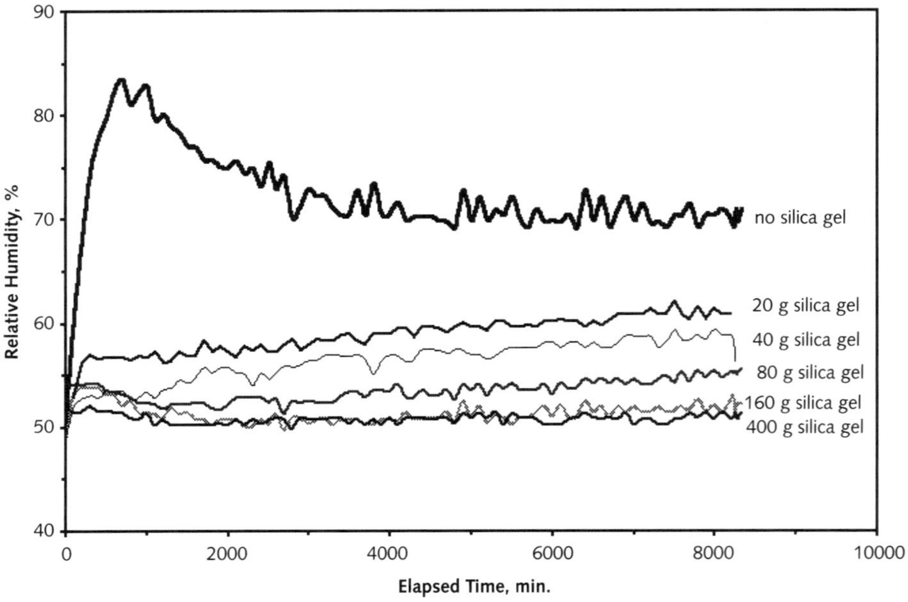


**Figure 3.4**
Relative humidity in 16 L plastic bags made of Filmpak 1193 and containing 4 packets of Ageless® Z-1000 with varying amounts (given as dry weight) of silica gel conditioned at 50% RH. About 20 g (dry weight) of silica gel per packet of Z-1000 were sufficient to limit the RH increase in the bags to 5%, that is, from 50% RH at the start to a final 55%.

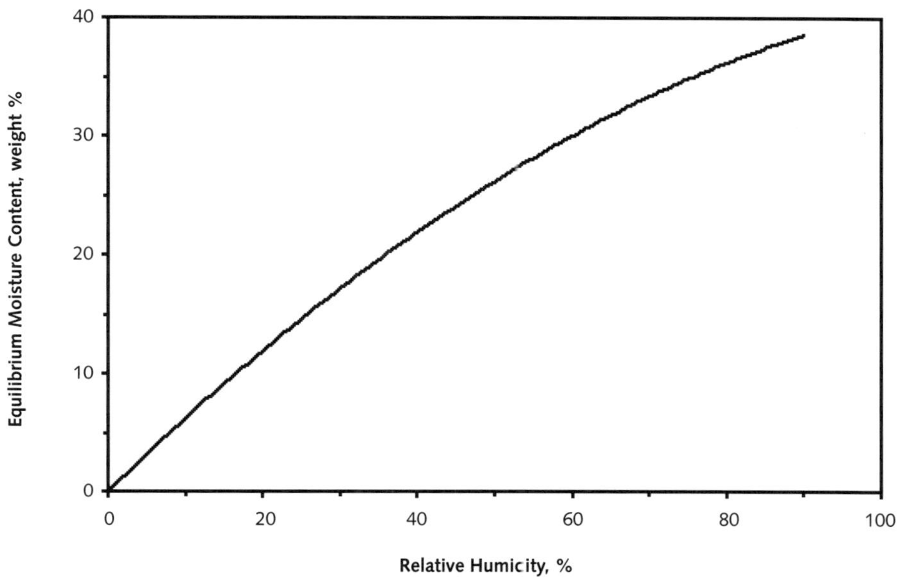


**Figure 3.5**
Sorption isotherm of bead-type silica gel

that is, from 50% to 55% RH. The line for 80 g of silica gel in Figure 3.4 provides support for this recommendation. The rate of absorption by this amount of silica gel was fast enough to maintain the relative humidity even in the first two hours of the experiment when the four packets of Ageless® Z-1000 released water very rapidly. (With objects also in the bag that had even a modest RH-buffering capacity, the change in relative humidity would be even more moderated.)

Figure 3.6 shows the results of experiments that were performed to find how well buffered silica gel would maintain a relative humidity lower than 50%. In the figure the sharp drop in relative humidity indicates when dry nitrogen was flushed through a 50 L bag

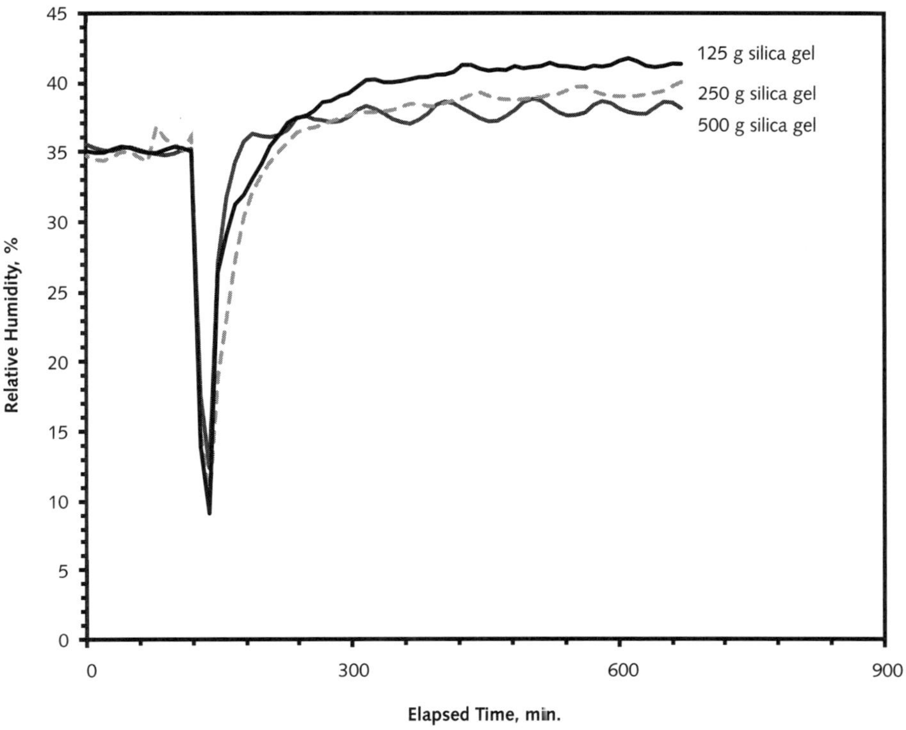


**Figure 3.6**
Plastic bags made of Filmpak 1193 with a volume of 50 L, containing 10 packets of Ageless® Z-500 and varying amounts (given as dry weight) of silica gel conditioned to 35% RH. Two hours after the start, the bags were purged with 400 L of dry nitrogen causing a radical drop in relative humidity. It took between 60 and 90 minutes for the RH in the bags to reattain the original value. Then, due to the moisture released by the oxygen absorber over time, the relative humidity stabilized at 37% to 42%, depending on the amount of silica gel present.

containing only silica gel conditioned at 35% RH plus Ageless® oxygen absorber that had a capacity of 5 L of oxygen. Three experiments were run with differing amounts of silica gel conditioned at 35% RH. The data show that within ninety minutes, about 30 g (dry weight, then conditioned to 35% RH) of silica gel for each Ageless® Z-1000 would restore the system to its original 35% RH. This ratio of silica gel to oxygen absorber will also limit the ultimate relative humidity increase in the bag to 5%, that is, an increase from 35% RH to a final 40%.

In Chapter 5 we show the simple method of calculating how much silica gel, conditioned to 50% RH, is needed for anoxia treatment enclosures to assure a stable RH of 50% to 55% for two weeks. The results are

Amount of RH-conditioned silica gel (in 100 g units) =
Amount of oxygen absorber (in 1000 cc packets) / 5.

In the example of the bag described in calculation 1 above for the amount of oxygen absorber needed for anoxia, 21 packets of 1000 cc capacity were required. Therefore, in that example, $2\frac{1}{5} \times 100$ g or 420 g of RH-conditioned silica gel would be the minimum.

## Notes

1. The properties of various types of MGC oxygen absorbers are shown in Table 3.1.
2. Oxygen absorber cannot be used for small anoxic treatments in everyday rigid plastic containers such as Tupperware because the internal pressure will be reduced during oxygen absorption, causing a deformation of the container and allowing air to leak in. Rigid containers can be used only if they are capable of maintaining a perfect seal under conditions of changing pressure. This can be achieved, of course, by having a flexible bellows attached to the container, as is done with large hermetically sealed museum display and storage cases, but impractical for anoxia in a kitchenware vessel.

## References

**Kondoh, Y.**
1995. Personal communication, June 16, 1995 (Mitsubishi Gas Chemical Co., Inc., Tokyo, Japan).

**Lambert, F., V. Daniel, and F. Preusser**
1992. The rate of absorption of oxygen by Ageless®: The utility of an oxygen scavenger in sealed cases. *Studies in Conservation* 37:267–74.

**MGC (Mitsubishi Gas Chemical Co.)**
1994. *Ageless®—An Oxygen Absorber, Preserving Product Purity, Integrity and Freshness.* (Product Information 8/1994.) Tokyo: Mitsubishi Gas Chemical Co.

1996. *RP System™ for Metal Artifacts: Guide Book.* (Product Information.) Tokyo: Mitsubishi Gas Chemical Co.

1997a. *RP System™: Protection against Oxidation of Metals and Electronic Parts—Stop the Clock!* (Product Information 3/1997.) Tokyo: Mitsubishi Gas Chemical Co.

1997b. *RP System™ For Cultural Artifacts: Usage Manual.* (Product Information 2/1997.) Tokyo: Mitsubishi Gas Chemical Co.

1997c. *Material Safety Data Sheet: RP Agent (Type A)*. (Product Information 3/97.) Tokyo: Mitsubishi Gas Chemical Co.

1997d. *Material Safety Data Sheet: RP Agent (Type K)*. (Product Information 11/97.) Tokyo: Mitsubishi Gas Chemical Co.

**Multisorb Technologies**
1996a. *FreshPax™ Oxygen Absorbing Packets*. (Product Information 9/96.) Buffalo, N.Y.: Multisorb Technologies, Inc.

1996b. *Material Safety Data Sheet: FreshMax, FreshPax D*. Buffalo, N.Y.: Multisorb Technologies, Inc.

**Nakamura, H., and J. Hoshino**
1983. Techniques for the preservation of food by employment of an oxygen absorber. In *Sanitation Control for Food Sterilizing Techniques*. Tokyo: Sanyu Publishing Co.

**Standa Industrie**
1994. *Atco® Oxygen Absorbers*. (Product Information 2/1994.) Caen, France: Standa Industrie.

2000. Atco Technical Specifications, on-line at www.atmosphere-controle.fr/specs.html (checked 4/14/00).

Chapter 4

# Large-Scale Anoxia Using External Nitrogen Sources

Anoxia can be carried out in large enclosures that range from user-made tents of up to 10 $m^3$, commercial flexible Bubbles that may be 50 $m^3$, and rigid steel chambers previously used in large museums for fumigation. (See Chapter 5 for step-by-step instructions for anoxic treatment on a large scale.) All of these containers for anoxic treatments require moderate to massive quantities of nitrogen gas to purge the oxygen from their atmospheres. Accordingly, after an initial survey of large systems and procedures for anoxia, we discuss the safety of operators who will be using nitrogen in these amounts. This is followed by descriptions of various sources of nitrogen, humidification systems, fabrication of tents, and operation of the several kinds of large enclosures used in anoxia.

## Overview

Figure 4.1 contains the features common to most large anoxic treatment systems. The nitrogen can be supplied by cylinders, tanks of liquid nitrogen (often called dewars, after their originator, Sir James Dewar), or nitrogen generators. Because the nitrogen gas from these sources is dry, it must be humidified to the relative humidity most suitable for the objects being treated before it flows into the flexible plastic film or rigid enclosure. Cubical tentlike enclosures can be readily constructed from plastic film using the methods already described for small bags, except that films 0.1 to 0.15 mm thick must be chosen so that a tent is durable enough for

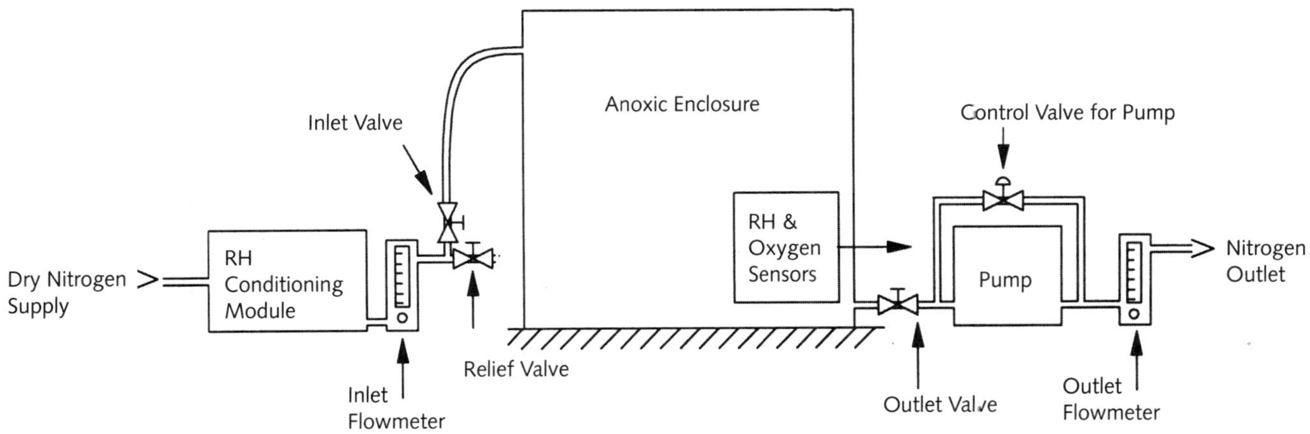


**Figure 4.1**
The large-scale treatment system consisting of nitrogen source, humidification system, enclosure for anoxia, vacuum pump, flowmeters, and monitors for humidity and oxygen concentration

reuse. Bubbles, large portable tents of oxygen-impermeable film, are available from the manufacturer, Power Plastics, in the United Kingdom. Some museums have metal chambers, previously used for fumigation with toxic gases, that can easily be converted to excellent enclosures for anoxia with nitrogen gas.

A distinguishing feature of large-scale anoxia is the requirement for accurate oxygen, RH, and temperature sensors inside the anoxic treatment vessel so that the exact state of its atmosphere can be known at all times. Radical changes in relative humidity can occur in a few hours and if not countered may harm the objects being treated. Oxygen changes may be slower, but action must be taken promptly to ensure that the oxygen level does not rise above 0.3% if the treatment is to be effective in eradicating 100% of all insect life stages in fourteen days. The vacuum pump and outlet flowmeter in Figure 4.1 are essential in the basic operations of anoxia on a large scale.

The overall procedure is as follows. If the enclosure for anoxia has been tested and found adequately leak-free and the humidification system for the nitrogen supply checked to confirm that it can produce 55% RH nitrogen gas (or another desirable RH level), the infested artifacts are placed in the enclosure or chamber. It is then flushed with humidified nitrogen until the oxygen concentration is reduced to less than 0.3%. (This normally requires a volume of nitrogen gas seven to eight times the volume of the enclosure.)[1] When that level is reached, the nitrogen flow can be reduced so that a minimum amount to maintain the oxygen below 0.3% passes into the treatment enclosure. The outlet port should be distant from (diagonally and higher than) the nitrogen inlet. This simple slow flow works well for a relatively low-leak enclosure for the duration of the anoxia process. A very leaky enclosure dramatically increases the need for nitrogen and increases the complexity of controlling the relative humidity as well as the possibility of rapid RH change. This can be surprisingly costly in nitrogen consumption in a large container and potentially hazardous to its contents. In contrast, extensive work at the GCI has proved that tents can be constructed with such low-leak qualities that after the initial oxygen level of about 0.1% is achieved by flushing with nitrogen, no additional nitrogen gas need be passed into the tent for the entire fourteen-day period.

Failure or drift of the humidification control over a long treatment period can cause disastrous hygroscopic stress on the objects. For example, if an enclosure requires one complete volume exchange with nitrogen per day to maintain the anoxic environment, drifting or failure of the humidity control may result in changes of up to one-half of the set RH value in a day. That is, the relative humidity could drop from 50% to 25% or rise to 75% within twenty-four hours. Clearly, the chances of RH instability increase as the purging rate has to be increased. Therefore, an airtight enclosure that requires the smallest purging rate with respect to its volume is least costly as well as least likely to pose any risk to the objects in it.

## Safe Use of Large Volumes of Nitrogen

Nitrogen is nonflammable, odorless, and nontoxic. Air is nearly 80% nitrogen. However, when more than that amount of nitrogen is introduced into the atmosphere in which a person is working, the decrease in oxygen content can result in impaired coordination and judgment and even unconsciousness (Air Products 2001a). Thus it is essential for safety that a person working with large quantities of nitrogen carry a pocket oxygen monitor with a built-in alarm. It is not that nitrogen itself is harmful, but with this protective warning one can be aware immediately that the concentration of oxygen is greatly decreased in the work area and avoid any serious consequences. Such a simple safety measure is therefore recommended to eliminate any risk of accidents when great quantities of nitrogen are used.

There is no hazard in venting 100 or 200 L of nitrogen gas over fifteen to thirty minutes in an ordinary room with forced-air ventilation when purging a plastic anoxia bag. (This assumes, of course, that one does not stay in the direct path of all the nitrogen exiting the bag and that normal mixing of nitrogen with room air is constantly occurring.) The safety cautions given here are pertinent to the release of 100 L *per minute* or more in a work space because it might not be diffused quickly and a person could be in the direct stream of an excessive nitrogen flow.

OSHA has determined that 19.5% oxygen is the allowable minimum concentration for confined spaces. Personnel should not enter areas where oxygen concentrations are below this level unless provided with a self-contained breathing apparatus (Air Products 2001a). In confined spaces it is necessary to install oxygen monitors with an alarm that sounds at the 19.5% oxygen level. Monitors should be located away from doors or fresh air inlets to ensure that they are giving a true response to the general atmosphere in the room and not one that is falsely optimistic. Oxygen monitors often operate on batteries with normally limited service life and therefore must be scheduled for regular battery replacement. As a precaution, those working directly with the nitrogen supplies should carry personal monitors. A routine OSHA requirement is the use of eye protection when working with compressed gases.

Often at the start of purging large enclosures, very high flow rates are used to quickly reduce the oxygen concentration. However, this may result in more than 100 L per minute of nitrogen gas exiting from the enclosure into the working area. This is how a hazardous situation may be created. Any such problem can readily be avoided by leading all the nitrogen coming from the tent to outside the building through a simple large-diameter duct. An alternative system but one that is more complex and less direct is a system that provides auxiliary fresh air to an unusually well ventilated room or a room that has windows that open. It is essential that individuals be accompanied by a colleague if personal oxygen monitors are not carried when working in a room where a large-scale anoxic treatment is performed. However, this is not the best policy. The prudent course is to wear a personal monitor at all times.

If liquid nitrogen is chosen as the nitrogen gas source in anoxia, using safety gloves and glasses is advisable. Although normally only nitrogen gas will be dealt with when using liquid nitrogen for anoxia (see below), the liquid might be encountered accidentally. Liquid nitrogen is very cold (−196°C), and skin coming into contact with it for more a second or two can be quickly frozen, possibly resulting in severe burns. Liquid nitrogen is not to be feared, simply treated with respect (Air Products 2001b). Additional safety information can be obtained from Material Safety Data Sheets that are available from local suppliers of gases and cryogenics.

## Nitrogen Sources: High-Pressure Gas, Liquid Nitrogen, Nitrogen Generation

Selecting the kind of nitrogen source to use depends on the size of the anoxia enclosure and its leak rate. A rule of thumb for relatively low leaking units is that nitrogen gas cylinders are best for enclosures up to 5 $m^3$ (5000 L) and liquid nitrogen tanks, or dewars, are best for units that are about 5 $m^3$ and larger. Nitrogen generators, costly but available on a long-term purchase-to-own plan, efficiently separate the oxygen from nitrogen in air. They are convenient pure nitrogen sources for - anoxia enclosures throughout a large range above 5 $m^3$. (Nitrogen generators require a storage tank to accumulate pure nitrogen for rapidly flushing chambers larger than 10 $m^3$.)

Prices for nitrogen cylinders, containers of liquid nitrogen, and large nitrogen generators vary from country to country, usually even locally, and the most suitable size chosen by the person in charge of anoxia might not be the least expensive in a particular city. Therefore, it is essential to obtain price quotes, including peripheral costs such as charges for cylinders or liquid containers, from several companies. (Some manufacturers that have local dealers all over the world are listed in Appendix IV.) Prices usually include delivery and pickup of cylinders and tanks. The bulky, very heavy dewars of liquid nitrogen, as well as cylinder transporters other than two-cylinder carts, are not easy to maneuver in narrow spaces. Thus easy access for cylinder and tank carts coming from the loading dock, as well as ventilation of the room in which large-scale anoxia is to be carried out, should be evaluated.

If a nitrogen generator is the best choice for the museum or archive, its size and weight must be considered, along with space for a pure-nitrogen storage tank (for cost-effective operation). Although smaller generator units are neatly mounted on a pallet, even this configuration requires a clear pathway for a forklift to maneuver easily from the loading dock to its final location near the anoxic treatment enclosure. In addition, the need for work space around the generator and its storage tank requires careful site planning. Nevertheless, the final result of carefree and effective production of large volumes of nitrogen makes a nitrogen generator advisable for any institution that will use anoxia

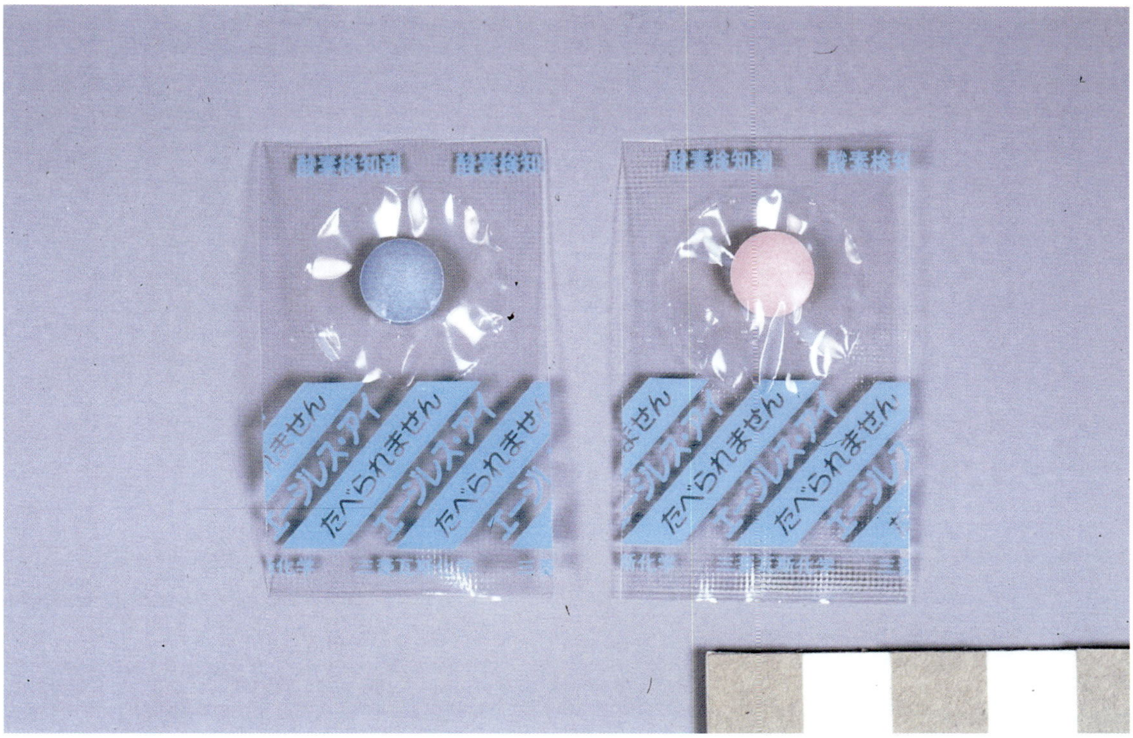


**Plate 1**

Left, Ageless-Eye® oxygen indicator after brief exposure to air (violet); and right, after a lengthy exposure in a low-oxygen (approx. 0.1%) atmosphere (pink)

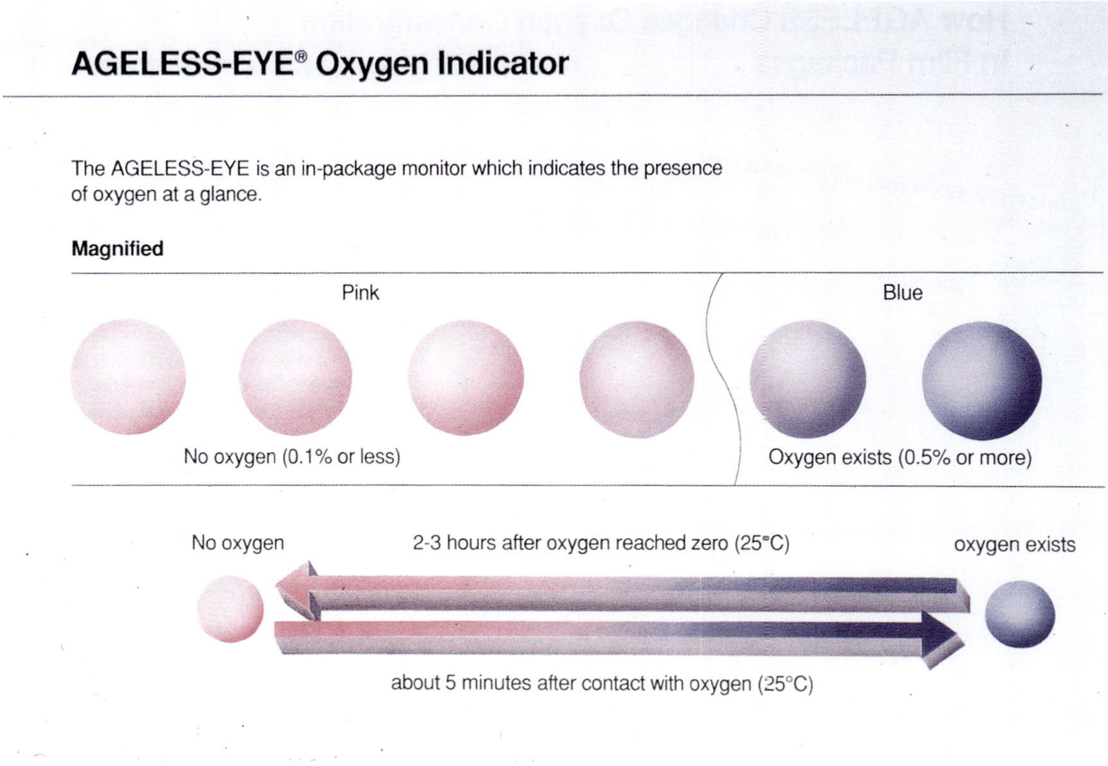


**Plate 2**

Commercial illustration of the performance of Ageless-Eye® oxygen indicator (MGC 1994)

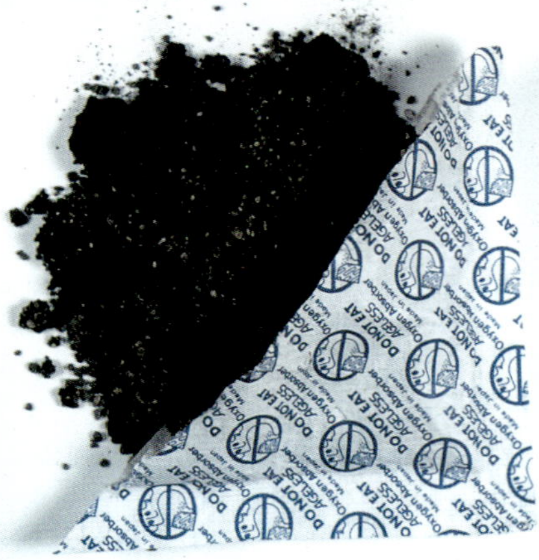


**Plate 3**

Most oxygen absorbers are an iron oxide powder that reacts with oxygen to form other iron oxides.

**Plate 4**

Several brands of oxygen absorbers are available on the market. The packets shown here have different absorption capacities (scale = 5 cm).

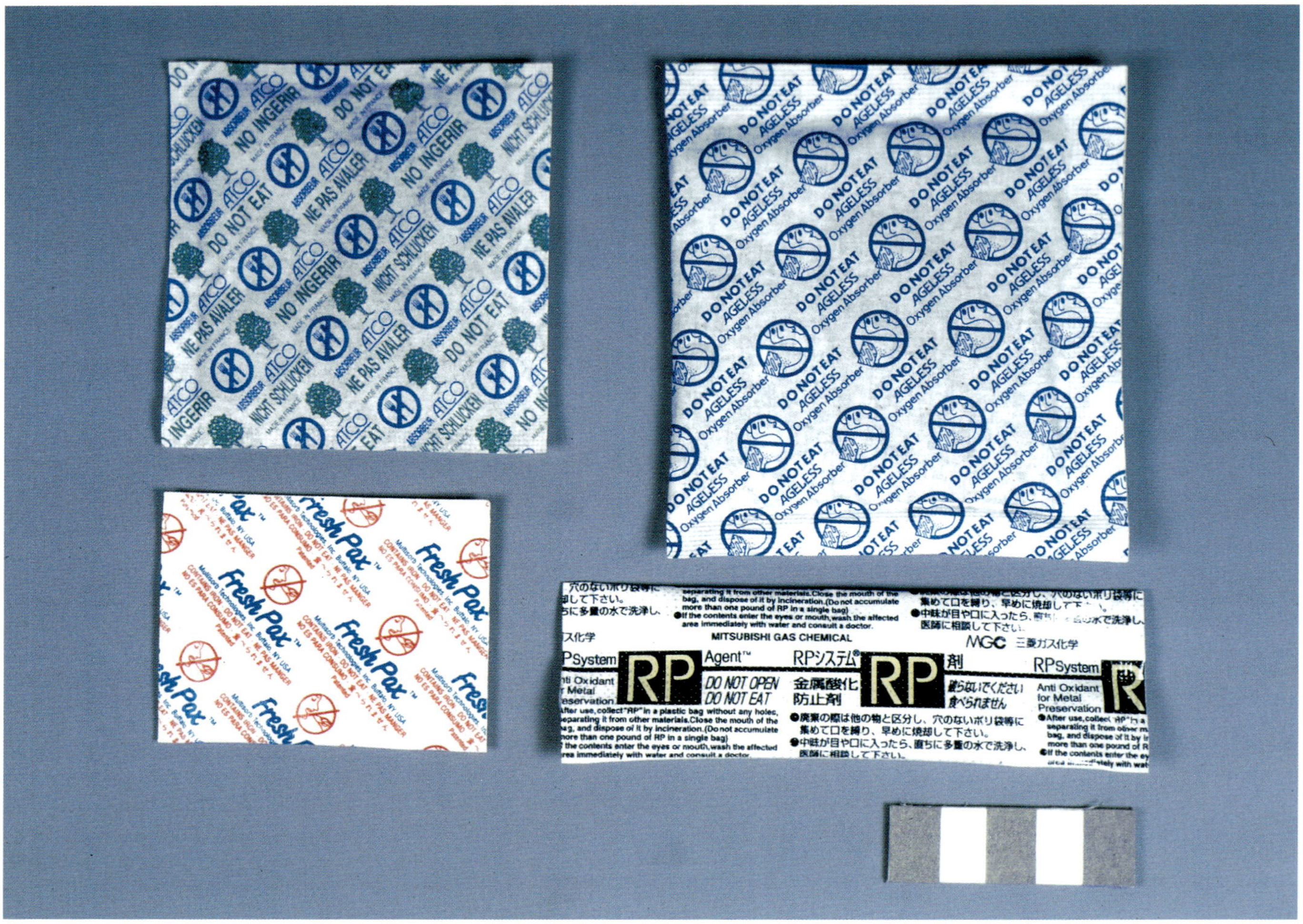

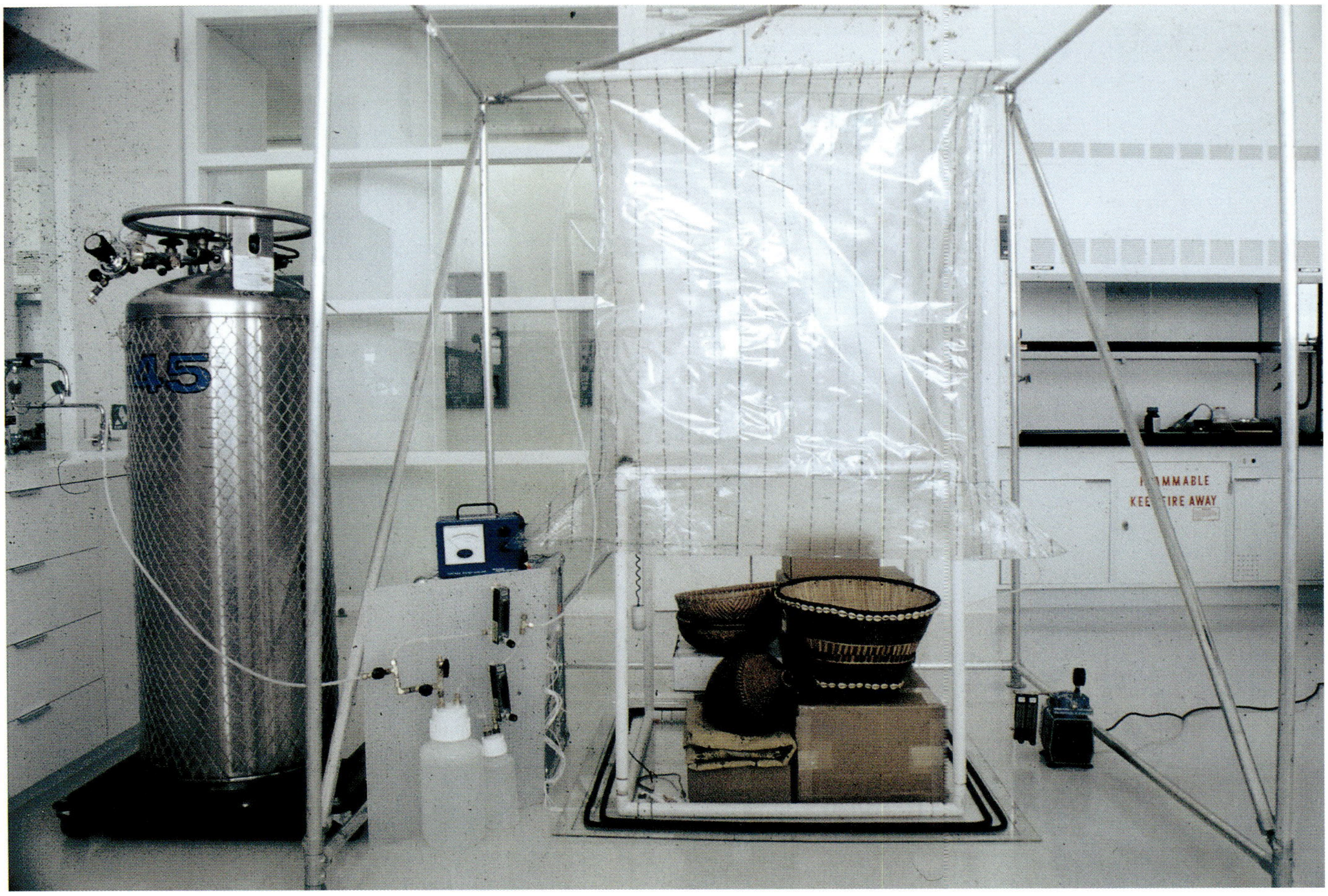


**Plate 5**

Completed 1 $m^3$ anoxic treatment system with accessories of liquid nitrogen source, nitrogen humidifier, and oxygen monitor. The tent has been lifted so that objects can be placed in it.

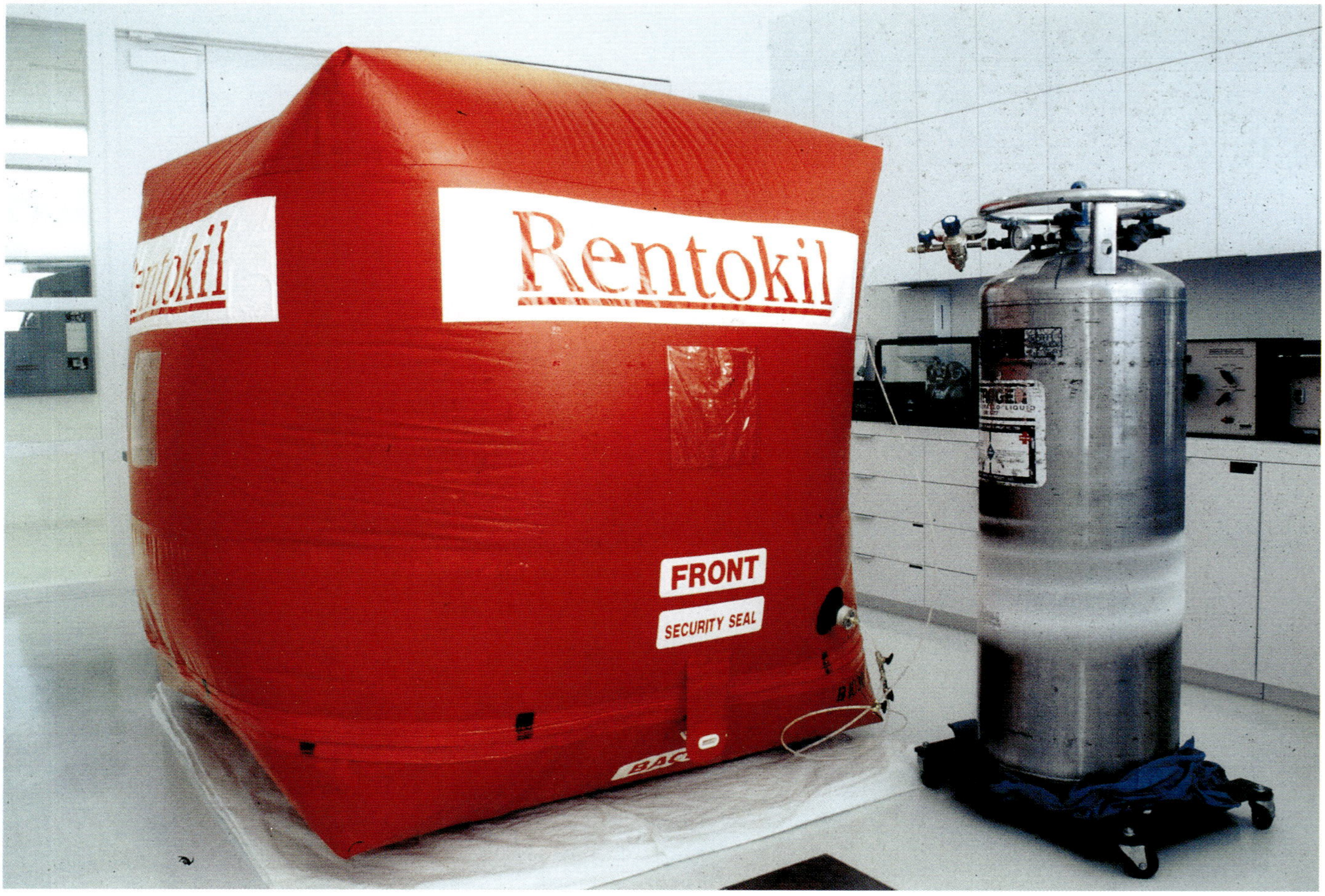


**Plate 6**

A 6 m$^3$ PVC Bubble was tested at the GCI using liquid nitrogen as the nitrogen source

for an extended period, if budgetary constraints do not prevent its purchase or rental.

### Nitrogen gas cylinders

In the past high-pressure (170 barg, 2250 psig) nitrogen cylinders, containing about 8600 L of gas at atmospheric pressure, were the principal nitrogen source for research in anoxia and for the anoxic treatment of infested objects. These cylinders, designated as T or 1 L size depending on the U.S. supplier, are about 140 cm high and 24 cm in diameter and weigh 73 kg when filled with 10 kg of nitrogen. They are very useful for purging smaller enclosures up to several thousand liters, that is, several cubic meters. To avoid the situation of emptying a cylinder before another one or two have arrived from the dealer, a number of cylinders are often lined up connected or connectable to a common manifold (Fig. 4.2).

Alternatively, for enclosures of several cubic meters, it is wise to use "portable banks" of nitrogen cylinders, units of six to twelve cylinders connected to a common manifold and mounted on a steel pallet (Fig. 4.3). (U.S. dealers in gases often call the six-cylinder units "six-packs." They, and larger packs, require a forklift to be moved.) Because each cylinder can provide approximately 8 $m^3$ of nitrogen at atmospheric pressure, an eight-cylinder "bank" can supply 64 $m^3$ of nitrogen, more than enough to flush a 5 $m^3$ enclosure eight times and maintain its anoxic atmosphere for fourteen days. If a forklift and adequate hallway space are available, these portable banks of cylinders are not only more convenient but also far safer than dealing with a number of individual cylinders. There are many potential hazards if a person must move six or eight heavy nitrogen cylinders, one by one as they are needed, from storage. Each must be secured in a special hand truck and attached to a gas fitting; then the empty must be detached and moved back to storage during the flushing and maintenance steps in anoxic treatments.

**Figure 4.2, below**
Nitrogen cylinders can be connected in parallel to decrease the need for frequent cylinder changes.

**Figure 4.3, below right**
Cylinder "cradles," or "six-packs" of 6 cylinders, or banks of 12 or more can be mounted on a mobile steel pallet and connected to a common manifold.

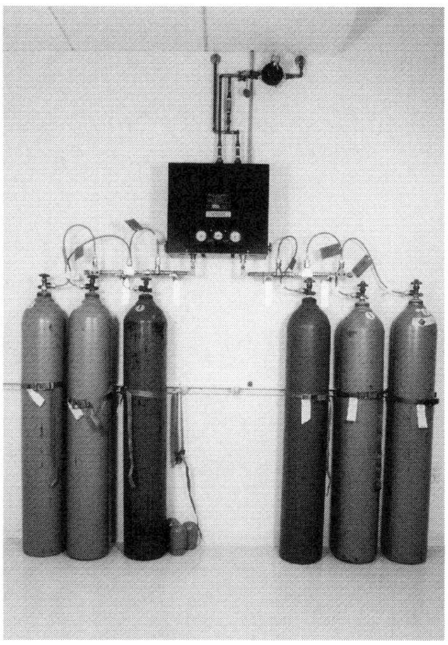

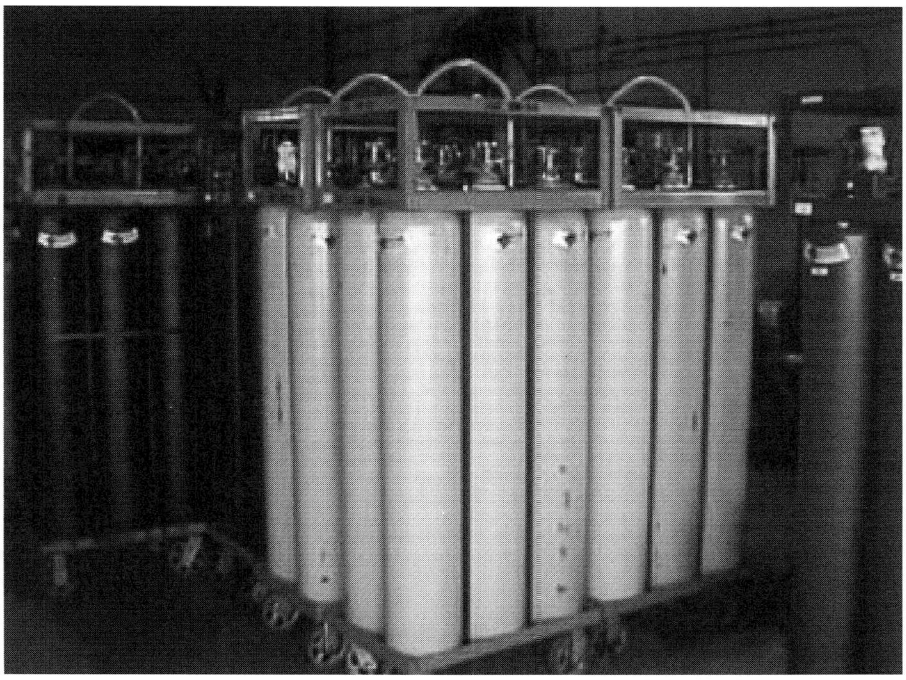

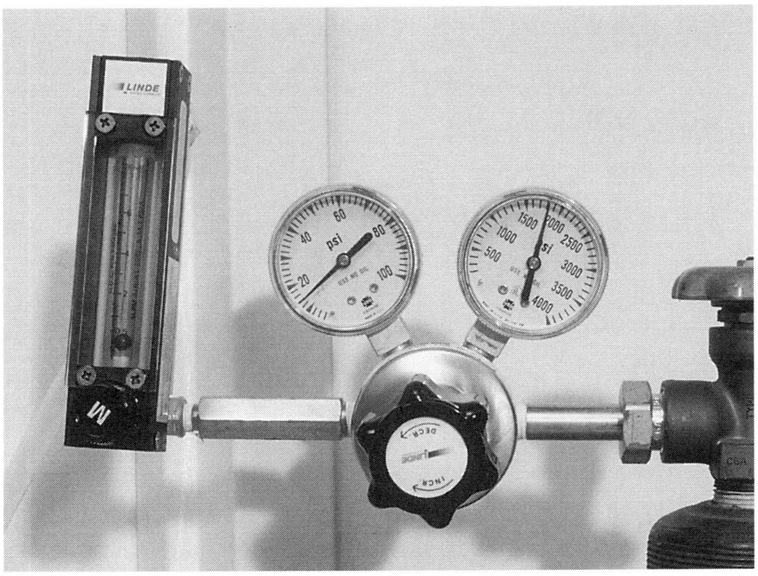

Figure 4.4

A regulator is essential to control and monitor the nitrogen pressure at its source. (The flow rate is shown by flowmeters in the humidification system and one located near the inlet of a large tent.)

Nitrogen gas comes in cylinders in various purities (grades). Industrial grade nitrogen, at about $24 per T cylinder, is the least expensive. However, it has limited use for anoxic treatments because its guaranteed nitrogen content is only 99.7%, and thus it could not practically achieve a 0.3% oxygen level. High-purity nitrogen (99.98%) with approximately 10 ppm of oxygen costs $35 to $45 per cylinder. It is the grade recommended for anoxia use. (Prepurified nitrogen has only 5 ppm of oxygen and thus is far more pure than needed for anoxia.) For all grades, the rental fee per month is in the range of $6 per cylinder after the first month. To reduce their pressure to about 3.5 barg (50 psig), nitrogen cylinders must be equipped with a regulator and gauges. A two-stage regulator, which costs about $240 (Fig. 4.4), is advisable because it maintains the selected pressure in spite of the decreasing cylinder pressure, thereby keeping the nitrogen flow rate constant. For safety, regulators have special fittings so that they can be used only for cylinders with a particular type of gas. In the United States a regulator for nitrogen has the CGA number 580.

## Liquid nitrogen containers

Liquid nitrogen is supplied in 100 L to 500 L stainless-steel insulated containers, or dewars, that are evacuated between the walls like a Thermos bottle. When liquid nitrogen strictly *for delivery of gas* is ordered, a special type of tank that contains an internal heat exchanger can be supplied (Fig. 4.5). This heat exchanger is necessary to warm the extremely cold liquid nitrogen (−196° C) and change it into nitrogen gas. One liter of liquid nitrogen is equivalent to about 700 L of gas at atmospheric pressure and 25°C. The gas pressure available from liquid nitrogen containers is not high, typically less than 15.5 barg (225 psig), but this is sufficient for most anoxia applications because a maximum flow rate of 120 L of gaseous nitrogen per minute can be produced. To control the flow, a standard two-stage regulator like that used with nitrogen gas cylinders is necessary. Higher flow rates can be achieved by using an external heat exchanger (a passive device) of the type recommended by the liquid nitrogen supplier.

A 180 L dewar, designated GP-45 by Praxair, is a convenient and widely used size, yielding about 103,000 L of gaseous nitrogen. (This quantity is almost the same as twelve T gas cylinders whose price would be about $480.) The purity of the nitrogen is 99.998%, with approximately 20 ppm of oxygen. Costs for the 180 L units range from $60 to $80 plus a $30 rental fee per month, after the first month. Inexpensive and needing no maintenance or mechanical components, one such tank of liquid nitrogen has more than the nitrogen gas capacity needed to purge

Figure 4.5

Liquid and gaseous nitrogen can be withdrawn from liquid nitrogen containers.

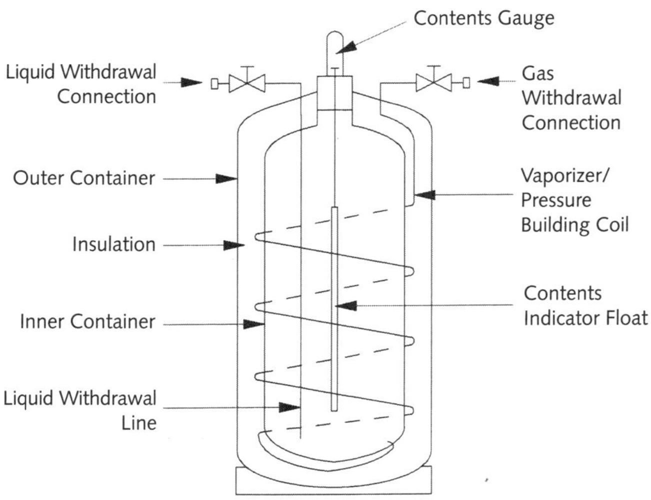

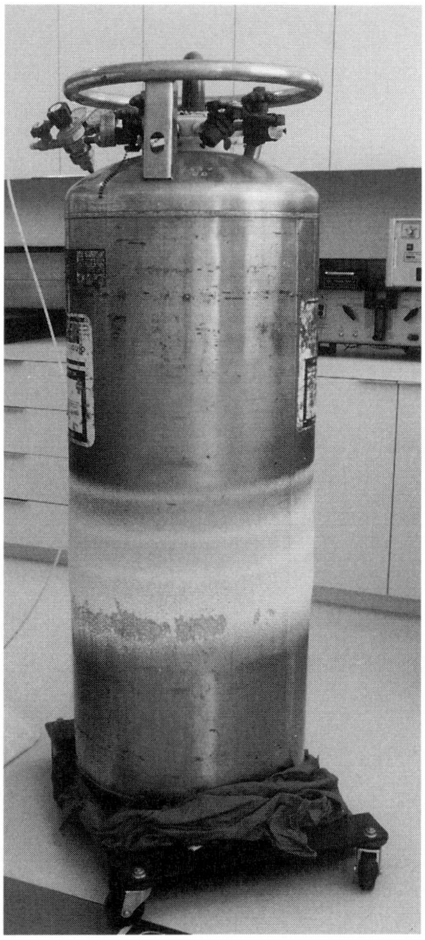

**Figure 4.6**

Ice will form on the liquid nitrogen container if high nitrogen flow rates are maintained over long periods.

a 10 $m^3$ anoxia enclosure to less than 0.1% oxygen and maintain that level for fourteen days. Somewhat inconvenient but not dangerous is the formation of large amounts of ice on the container (Fig. 4.6) when relatively high flow rates are continued for a long time. Under these conditions, precautions should be taken to collect the water that results when the ice melts so that a messy puddle that can pose a slippery hazard will be avoided. The only constraint on the use of liquid nitrogen is its "shelf life." Even when it is not used, a tank only lasts about one month because nitrogen gas must be continuously and automatically released to avoid excessive pressure buildup in the tank at room temperature.

## Nitrogen generators

Three commercial methods can produce very pure nitrogen (up to 99.99%) directly from air in small to large quantities per day.

Pressure swing adsorption (PSA) systems operate on the principle of preferential, that is, selective, adsorption of oxygen on a molecular sieve from compressed air. (The oxygen is released by the sieve when the pressure is reduced.) Air is compressed, filtered and dried, and passed alternately through two molecular sieve beds. When one bed is saturated with oxygen, it is depressurized to release adsorbed oxygen, carbon dioxide, and water vapor while the airflow is diverted to the second bed (Kirk et al. 1996). Commercially available PSA systems typically produce nitrogen at a rate of up to 50 $m^3$/min at purities between 95 and 99.9% nitrogen (Air Products 2001c).

The second method of separation, the high purity nitrogen (HPN) system, is essentially a local or in-house liquid nitrogen producer from air. Compressed, cooled air is liquefied by the same process used in multiton lots daily in industry and then, by distillation, separated into nitrogen and oxygen gas because of their different boiling points. HPN systems can deliver high-purity nitrogen at flow rates of more than 20 $m^3$/min, containing only about 5 ppm of oxygen (Air Products 2001d).

Selective gas permeation through membranes is the principle of a third major method for the preparation of nitrogen from air (Fig. 4.7). In these commercial systems air is compressed and filtered before it is fed into the membrane consisting of bundles of thousands of tiny hollow fibers. Oxygen, as well as carbon dioxide and water vapor, permeates faster through the fiber walls than does nitrogen, and thus the separation of nitrogen occurs. Totally automated membrane systems are available with a wide range of capacities, starting at only 1 L up to some 30 $m^3$ per minute, and with purities between 95 and 99.9% nitrogen (Air Products 2001d, 2001e).

### *Criteria for choosing nitrogen generators for anoxia*

Of the three varieties of nitrogen generators, the membrane system is most suitable for large-scale anoxia in museums. Generally, the capacities of the other generator systems, PSA and HPN, are far larger than required and costs are proportionately high. Furthermore, PSA generators require frequent maintenance and the very high purity produced by HPN units is not necessary for anoxia of insects in museum objects.

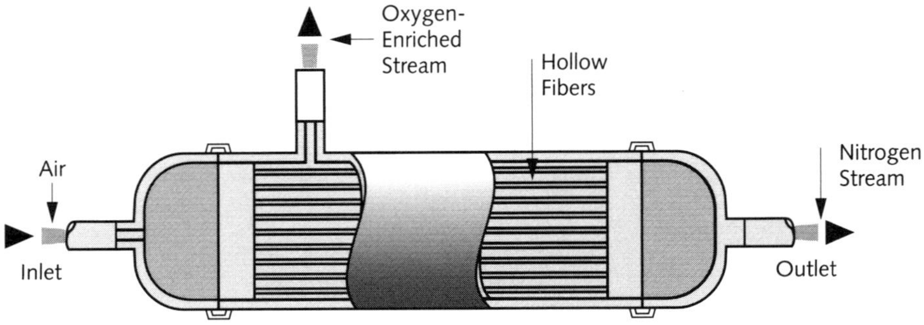


**Figure 4.7**
The principle of the nitrogen generator that produces pure nitrogen from air. Certain polymeric membranes have a selective permeability for different gases and separate nitrogen from oxygen because the oxygen moves through them more rapidly.

Membrane nitrogen generators, such as the PRISM® Nitrogen Membrane System (Fig. 4.8) manufactured by Air Products and Chemicals, are available as turnkey systems in many countries. This instantly ready—and permanently in-house—supply of nitrogen commends itself to a person with the following needs.

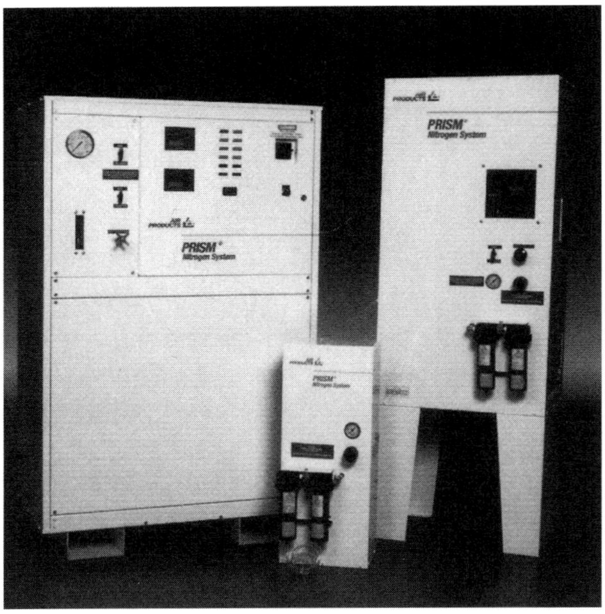

**Figure 4.8**
Nitrogen membrane generators are available in various sizes. The largest shown here has a capacity of more than 40 L per minute of 99.9% nitrogen.

*Personnel Requirements*

- Convenience and safety. Nitrogen immediately available on site at any time for a single anoxic treatment or an indefinitely extended series. No handling of heavy cylinders or tanks or concern for their safe storage.
- Economy. Labor saving: elimination of tanks or cylinders. Time saving: no need to order nitrogen, check its arrival, or continually monitor status of remaining nitrogen on hand to avoid running out during an anoxia treatment. Minimal cost for nitrogen, especially important when 1000 $m^3$ or more per year are used.

*Facility Capabilities*

- Anoxic enclosure size. Optimally, permanent 5 $m^3$ or larger unit (to justify capital cost of membrane nitrogen generator system).
- Space available. Access to the room with the anoxic enclosure and placement there of the membrane system that has a base size of 1 m × 1.5 m. Additional floor or wall space needed for cylindrical nitrogen storage tank.
- Utility requirements. 120 VAC 1-Phase (generator), 220 VAC or 230 VAC 3-phase (compressor).

The smallest fully automatic PRISM® nitrogen generator suitable for large-scale anoxia, Model 1100, can deliver 14 $m^3$ of 99.9% nitrogen per day (or 34 $m^3$ of 99.0% per day). A Model 1300, with 41 $m^3$ per day of 99.9% nitrogen capacity, was available for research by the GCI at the time its 10 $m^3$ anoxia tent had been fabricated.

The PRISM® 1300 nitrogen generator, like other models, consisted of an air compressor and a series of filters, followed by the membrane separating units, and was equipped with an oxygen analyzer for the output gas. Its operation involved compressing intake air to 5.4 to 12.9 barg (80–190 psig) and passing it through prefilters to remove moisture, dust particles, and any gaseous pollutants before it reached the membranes. There, oxygen was separated as a side stream. As the manufacturer had guaranteed, the system was capable of producing nitrogen of 95% to 99.9% purity, inversely dependent on the flow rate. The maximum achievable flow rate was 28.5 L per minute at a purity of 99.9% nitrogen, but it was possible to improve the purity further to about 99.99% nitrogen by reducing the flow rate to approximately 15 L per minute.

The price, approximately $15,000, for the unit with a 230 VAC 3-phase motor/compressor is high for serving a 10 $m^3$ anoxic enclosure because the generator's capacity of 41 $m^3$ of 99.9% nitrogen per day is more than is needed. (At approximately $13,000, the Model 1100 with a capacity of 14 $m^3$ of 99.9% nitrogen and 31 $m^3$ of 99% nitrogen per day would be somewhat more cost appropriate for a 10 $m^3$ tent size.) These nitrogen generators are available for purchase-to-own on a five-year lease for roughly $350 per month, an arrangement that includes regular maintenance. The GCI recommends the comparison of cost-effectiveness of other nitrogen sources (including labor cost) versus purchase or rental of this kind of system. Convenience cannot be quantified, but it is an important consideration in choosing a membrane nitrogen source over cylinders of gas or tanks of liquid. The generator's operational cost, other than power consumption of the compressor, would amount to about $725 per year for replacement of prefilter cartridges and the oxygen sensor in the analyzer.

The PRISM® nitrogen generator was easy to operate and completely reliable for the two and a half months of nonstop use during the test.

## Humidification of Nitrogen

Nitrogen—whether from gas cylinders, liquid nitrogen, or membrane filter generators—contains only a few parts per million of water. Therefore, it must be humidified before it enters the anoxic enclosure to avoid changes in the moisture content of objects being treated.

In earlier work, a simple humidification system consisting of three bottles of an equal volume was used (Fig. 4.9). One bottle contained water through which a stream of nitrogen was bubbled to yield a nearly saturated gas flow. This was mixed with a flow of dry nitrogen in the second bottle, where any water droplets were removed, and the relative humidity of the mixed stream was measured in the third bottle just prior to passing into the bag or other enclosure for anoxia. Control of the relative humidity was accomplished by adjusting the flow ratio of the dry and wet streams based on the RH reading in the third bottle. Although the system was easy to assemble, the disadvantages were that, because the bottles were pressurized, the RH reading in the third bottle was always higher than that at atmospheric pressure and the system responded slowly at low nitrogen flow rates due to its large total volume. This made exact control difficult, especially since the RH sensors then used took about five minutes to equilibrate.

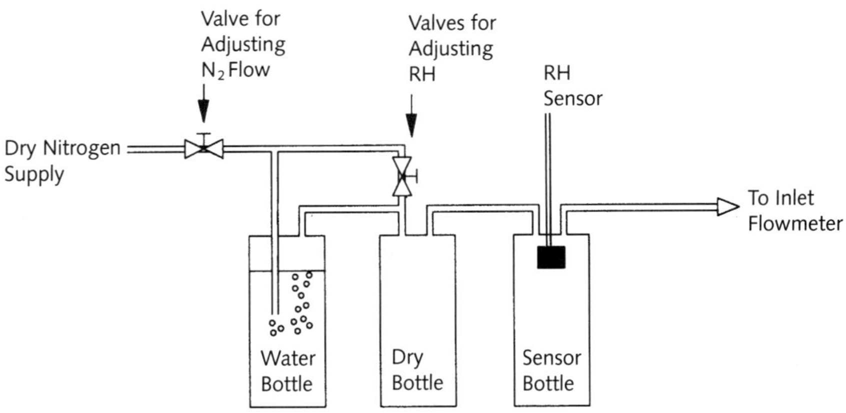


**Figure 4.9**
Previous version of the humidification system for dry nitrogen

An improved humidification system was developed that can be used in any size of anoxic treatment enclosure. First, a major modification was made by inserting flowmeters (with built-in needle valves) in the wet and dry nitrogen streams to ensure more precise and predictable control of those nitrogen flows (Figs. 4.10–4.12). Second, the system's total volume was reduced by eliminating the third bottle and reducing the volume of the second (droplet collector) bottle. For the initial calibration of the apparatus, a fast-acting RH sensor was placed in a plastic exit bag that was kept at atmospheric pressure by allowing unrestricted gas flow out of it. These changes improved the system's response to RH changes and the accuracy of RH measurement during calibration and assured a supply of nitrogen with known relative humidity to the large enclosure when the calibration bag was replaced by a direct connection to the enclosure.

Theoretically, an equal flow in each of the wet and dry nitrogen gas streams (as shown by their respective flowmeters) should

**Figure 4.10**
Schematic of the improved version of the humidification system

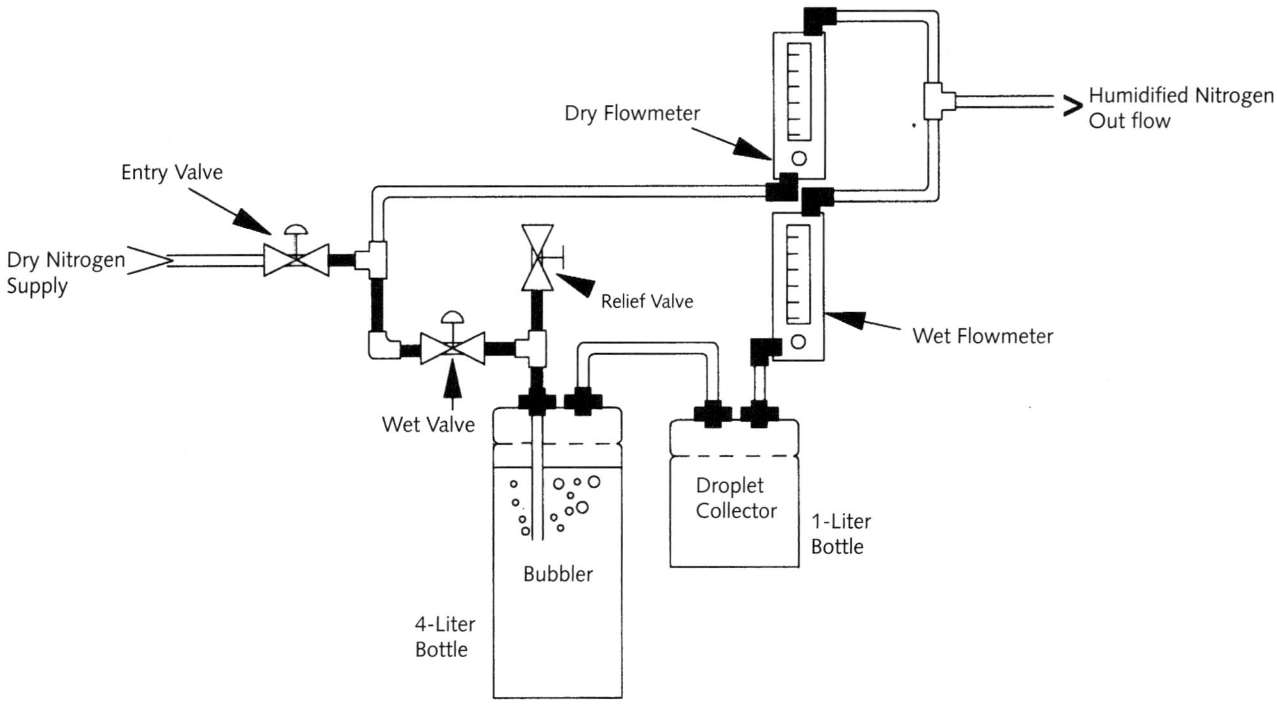

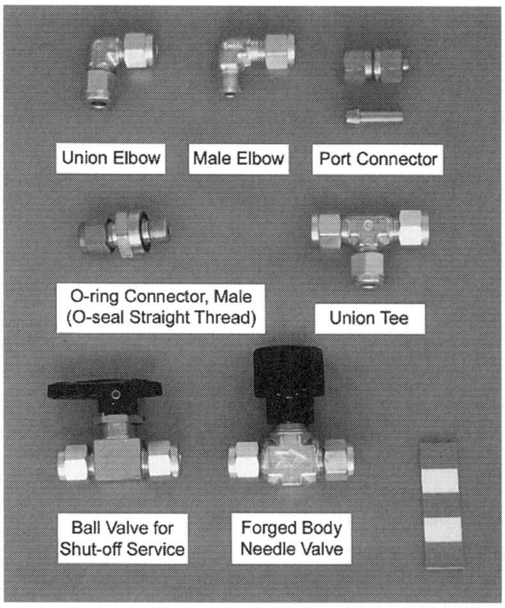


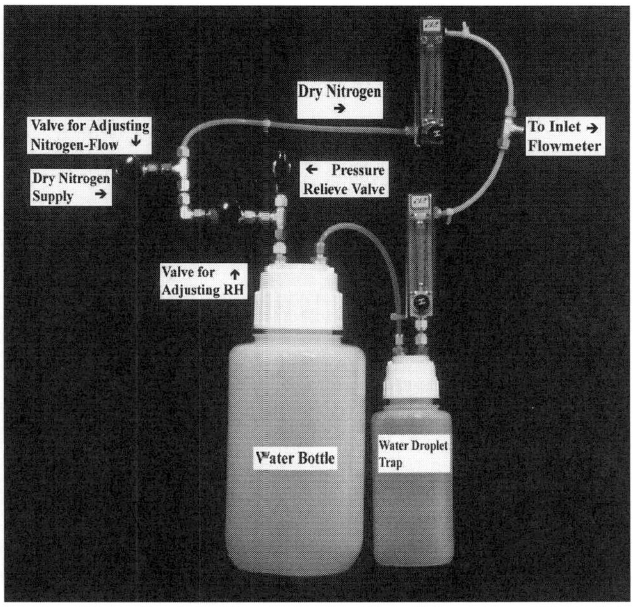


**Figure 4.11, above**
Valves, O-ring connectors, and other tubing fittings for joining the parts of the nitrogen humidifier

**Figure 4.12, above right**
The completed humidification unit

produce nitrogen with 50% RH. However, at high flow rates of the wet stream (even with the dry flowmeter valve turned off), it is not possible to fully saturate the nitrogen as it bubbles very rapidly through water at room temperature. Thus it is necessary by trial-and-error variation of the needle valve setting on the dry flowmeter and the needle valve leading to the water bottle to establish the exact wet and dry flow rates that produce a chosen RH value in the exit nitrogen at any given total flow. The optimal dry flowmeter setting (and wet flow needle-valve setting) can then be used in all future anoxic treatments.

If the humidification system just described is made using only a 4 L bottle as the wet stream humidifier, it is still capable of producing up to 30 L per minute of nitrogen of 50% RH. This is sufficient to permit flushing a 10 $m^3$ enclosure eight times its volume and reach a 0.1% oxygen level within about forty-five hours. (Using two humidification units in parallel, or heating the water bottle with a small commercial heating pad, would increase the humidification capacity and thus reduce the time required to flush a large enclosure.)

Step-by-step instructions for the construction and operation of the humidification apparatus are given in Chapter 5. A Spanish supplier of materials and equipment for anoxia, Metode, sells a humidification unit like this. Manufacturers of more expensive automatic humidifiers for nitrogen are listed in Appendix IV.

## Fabrication, Testing, and Use of Large Flexible Anoxic Enclosures

Large-scale anoxia treatments can be carried out in reusable tents or in large bags that are single use or have limited reuse. This section provides a general discussion of the design, construction, testing, and use of a 10 $m^3$ cubical enclosure that can readily be made by a conservator and used many times (Maekawa and Elert 1996). (See Chapter 5 for detailed instructions.)

Reusable tents should be made of a durable plastic film that will be unlikely to develop pinholes or voids when it is worked with or folded, such as Filmpak 1193, a laminated plastic film that is very impermeable to oxygen. Generally, plastic films with a thickness of 0.1 to 0.15 mm are preferable for large anoxia enclosures because they have to withstand the rigors of being handled many times and thereby being pulled, pushed, or wrinkled. Large bags, especially those that are specially made for a single anoxia treatment of one object, need not be so tough and resistant to use as Filmpak 1193. Therefore, they can be made of less expensive metallized film such as Marvelseal 360. The difference in cost can be considerable; for example, for the 10 $m^3$ tent described below, made from about 40 $m^2$ of film, the Filmpak 1193 cost \$300, and the Marvelseal 360 would be \$80.

The general process for the fabrication of bags is described in Chapter 2, and step-by-step instructions for small to medium-sized bags are given in Chapter 5. Large bags are more complex, so they take considerably longer to make (see Chapter 5), even though the overall difficulty of fabrication is not much greater. There are several reasons for the length of the process. The heat-sealing steps are extensive because there are many pieces that must be joined, not just three sides of a rectangle. A rigid acrylic base (or thoroughly recoated smooth concrete floor to avoid air leaking up through the concrete) and pulley system (to raise the tent for entry to the space beneath it) have to be constructed. An inner cube of polyethylene pipe must be built to protect the objects being treated. Finally, holes have to be carefully made in the tent for connectors for nitrogen entrance and exit (and, optionally, for the sealed connector between wires from sensors inside the tent to wires leading to detectors or recorders outside it).

Reusable tents of any desired size and shape can be fabricated by heat sealing wide strips of plastic barrier film. Practical experience showed that cubical or rectangular boxlike enclosures can be manufactured relatively easily, using only a manual heat sealer. A nearly cubical test tent of 10 $m^3$ volume, with a base of 2 m × 2 m and 2.5 m in height, was fabricated at the Getty Conservation Institute (Fig. 4.13). Construction of the tent from 40 $m^2$ of plastic barrier film required approximately sixteen person hours and would have cost about \$600 if inexpensive PVC pipe and sandbags had been used—as is recommended in the following highlights of design and construction.

To ensure wide contact and ultimately a superior seal with its base, the tent had a 20 cm wide extension or flange on each of its sides (Fig. 4.14). Stout loops were formed along the top edges to attach a lifting device, a square framework of aluminum tubing and fittings that makes it possible to raise the tent for easy placement of objects in it. (PVC pipe for this lifting unit is more flexible than aluminum tubing and thus less desirable.) For convenient raising, the device should be attached to a rope run through a pulley system at the ceiling (see Fig. 4.14; Plate 5 shows a 1 $m^3$ unit). A cubical framework on the floor is necessary to give internal support to the tent and thereby prevent any damage to objects inside it when the tent is lifted or lowered. PVC piping for this frame-

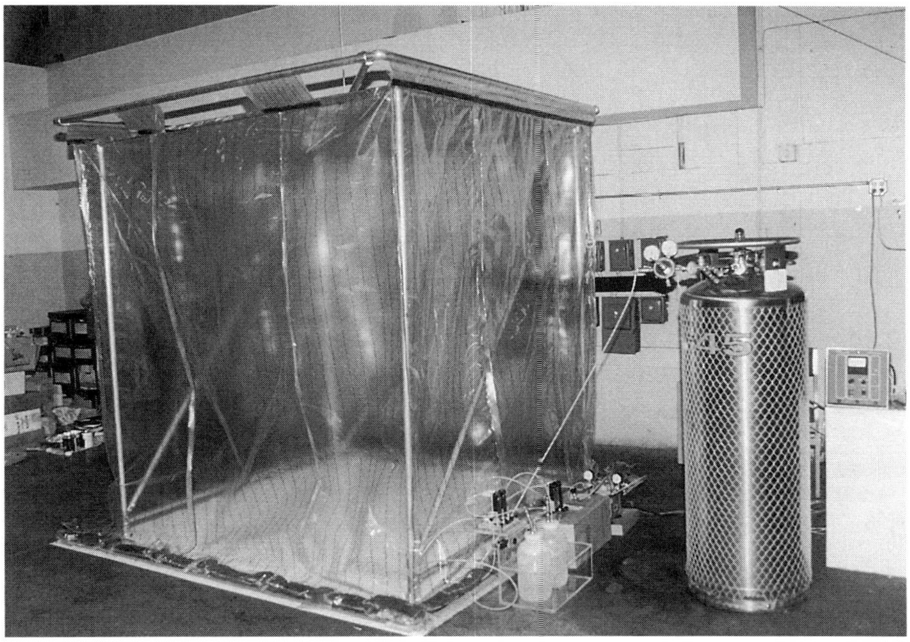

**Figure 4.13**
A 10 m$^3$ tent under test in a non-air-conditioned warehouse, liquid nitrogen humidifier on the floor, and beside it the vacuum pump (attached to outlet of tent)

work of a 10 m$^3$ enclosure should have a diameter of 2.5 cm, or, as shown in Figure 5.11 (p. 99), 2.0 cm diameter for a 1 m$^3$ tent.

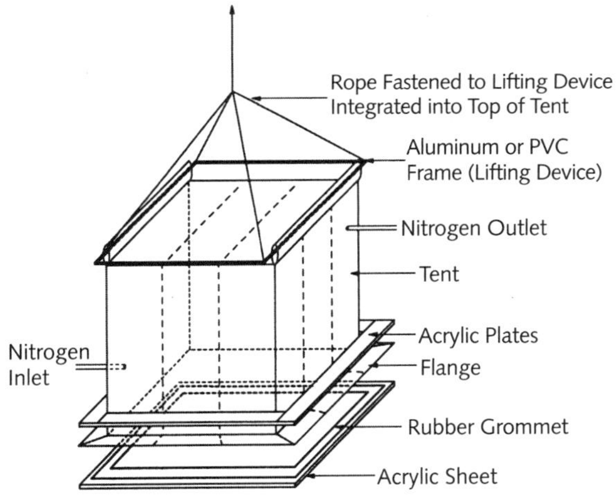


**Figure 4.14**
The tent cover has a wide flange to provide a good seal with the base and also has stout loops through which piping can be inserted as part of a lifting device.

Achieving a perfect seal between the flange of the tent and its base is an especially important goal for anyone making a tent. The base can be as simple as a smooth and crack-free concrete floor coated with several layers of epoxy paint (to prevent air from coming up into the tent) or a large acrylic panel at least 6 mm thick. A successful seal between base and flange is produced as follows (Fig. 4.15). Two self-adhesive rubber strips are placed in parallel lines marked on the floor beneath the flange perimeter. (Corner or other joints must not have any visible space between them. A razor blade is advisable for the cleanest possible cuts.) Then, after these strips are lightly greased with a laboratory grease or Vaseline, the tent is lowered and its flange carefully smoothed and adjusted to be wrinkle free. Acrylic plates, 1 cm thick by 15 cm wide, are placed on the flange perimeter. Finally, the flange is pressed tightly to the rubber strips by "sand snakes," long cylindrical weighting bags filled with sand, laid on the acrylic plates around the flange.

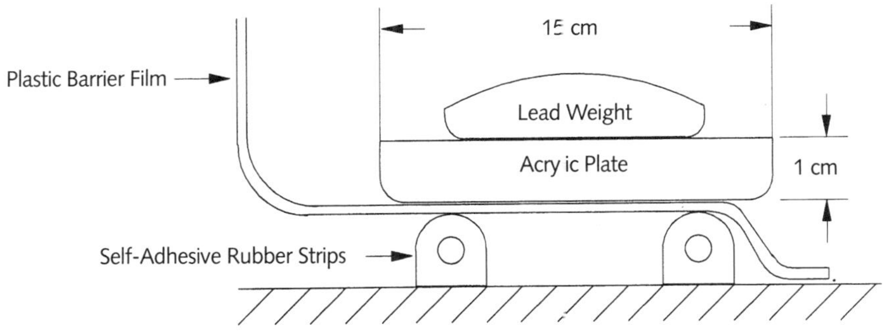


**Figure 4.15**
Lead shot weights on the acrylic plate press the flange of the tent cover tightly to adhesive rubber strips to provide a good seal. (Less expensive "sand snakes," long cylindrical bags filled with sand, are adequate substitutes.)

(Fig. 4.16 shows the use of lead shot, effective but far more expensive than sand snakes.) An adequate seal cannot be obtained by simply taping the flange to the floor or to a base sheet. Oxygen leak tests performed using tents of 1 $m^3$ volume revealed that the application of laboratory or vacuum grease or Vaseline on the rubber strips was essential because it reduced oxygen leakage through the seal by up to 80%.

Tests of the 10 $m^3$ anoxia tent were made in a non-air-conditioned warehouse and showed exceptional airtightness, that is, an oxygen leak rate of only 0.004% (40 ppm) per day after it was purged to an oxygen concentration of 0.1% (1000 ppm). Because of this well-sealed condition, the oxygen concentration after fourteen days reached only 1560 ppm (0.156%) even under static conditions, that is, with no flushing of nitrogen gas for fourteen days. These test results in a nonideal work space indicate that maintenance requirements can be minimal during an anoxia treatment if the enclosure is made carefully enough to ensure no major leakage.

Even if the oxygen ingress to a treatment tent is large enough to exceed the required anoxic level in a day or two, anoxia can still be successful; it merely requires additional nitrogen purging during the treatment. The only liability is the relatively large amount of humidified nitrogen that may have to be supplied for daily or continual purging. (See the description of an older 6 $m^3$ Bubble in the next section.) To ensure that an excess of nitrogen is not allowed into the tent, causing it to overexpand, the outflow must be exactly balanced with the input. This is readily achieved by having flowmeters on the inlet and outlet nitrogen lines and the outflow regulated by a vacuum pump whose suction is variable. (The process is described in detail in Chapter 5; for a schematic, see Fig. 4.1, p. 59.) Another method is to increase the size of the outflow opening until the inflation of the tent is controlled. The latter method is not recommended because it requires iterative adjustments and can fail too readily.

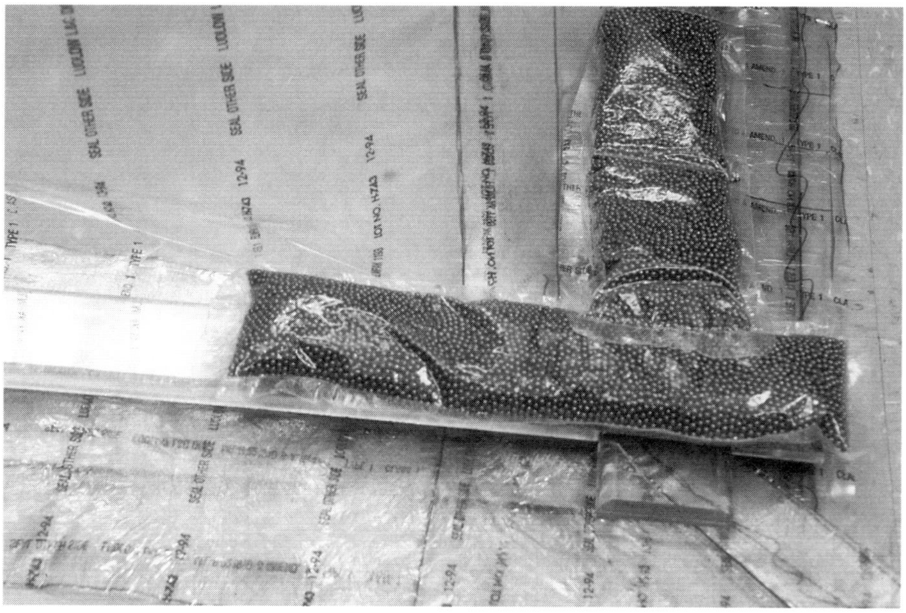

**Figure 4.16**

Acrylic plates and lead shot in plastic bags firmly seal the plastic tent cover to greased rubber strips.

## Tests of a Commercial Fumigation Bubble for Museum Anoxia

Large premade enclosures are available from Power Plastics in North Yorkshire, U.K., in sizes between 1 and 27 $m^3$. The bags are called Controlled Atmosphere Technology (CAT) Bubbles and are supplied as a top plus a base sheet (which the purchaser seals together) made from heat-sealable aluminized barrier film (Fig. 4.17; this Bubble was made by Power Plastics for Rentokil, Ltd.).

CAT Bubbles are reasonably priced, about \$110 for a 1 $m^3$ Bubble and \$250 for one of 27 $m^3$, but they cannot be reused. Provided with nitrogen inlet and exhaust and through-the-wall connection for an oxygen sensor, Bubbles do not have an internal humidity sensor. The oxygen monitor, hand-held heat sealer, nitrogen humidifier, and vacuum pump are sold as a "starter kit" (usable for all future Bubbles) at about \$5,500.

Although Power Plastics does not recommend it for use with nitrogen anoxia, the company also sells a PVC Bubble made from heavy, opaque plastic film that is available in a broad range of sizes, from 3 $m^3$ to larger than 50 $m^3$. It consists of a cubical dome-shaped top cover that can be attached to the base sheet by means of a 6 cm wide zip-lock, intended to provide an airtight seal. Windows made of transparent plastic film on opposite sides of the top cover can be ordered; they allow one to see the condition of an object during museum anoxia. Injection and exhaust ports are located on the top and side walls for attaching external gas supplies. Although the plastic material used in its construction is heavy, a Bubble can readily be folded and stored in a relatively small space when not in use.

This type of PVC Bubble has been used as a portable fumigation enclosure (Plate 6) for industrial commodities as well as museum objects by Rentokil Ltd. (West Sussex, U.K.). At present, Rentokil's practice is to use only Bubbles under its control. In the past conservators experienced problems regarding the airtightness of the PVC Bubbles then

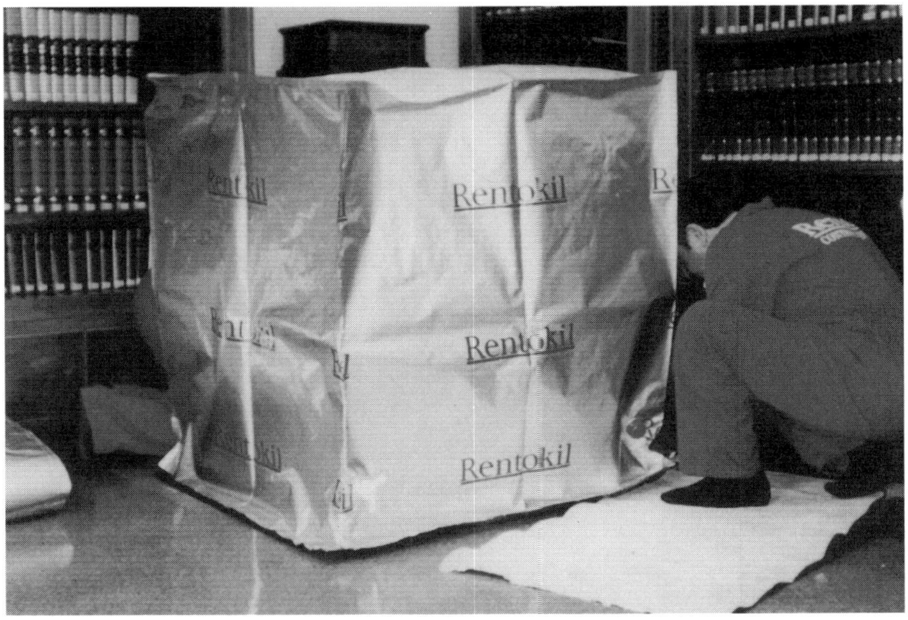


**Figure 4.17**
CAT Bubbles, large premade enclosures manufactured by Power Plastics (and used by Rentokil in their fumigation service)

sold by Rentokil (and now available only from Power Plastics), and therefore they refrained from using them with nitrogen. However, commercially available Bubbles are a convenient alternative to self-made large enclosures for anoxia, and a 1993 version of a Bubble was therefore tested at the GCI.

The 6 $m^3$ PVC-coated polyester Bubble, 1.8 m square on its base and approximately 1.7 m high, had two ports, one for the attachment of a gas supply and the other for a gas exhaust. Some minor modifications were necessary to adapt the Bubble for GCI use with nitrogen. Swagelok O-ring connectors that are preferred for nitrogen supply and exhaust were inserted in the plastic caps of the Rentokil ports after drilling. (Alternatively, custom-made caps for the O-ring connector can be fabricated from acrylic or metal to replace the port caps.) The Bubble also had to be fitted with a hermetically sealed electrical connector (shown in Fig. 4.18) for the installation of the necessary oxygen and RH sensors.

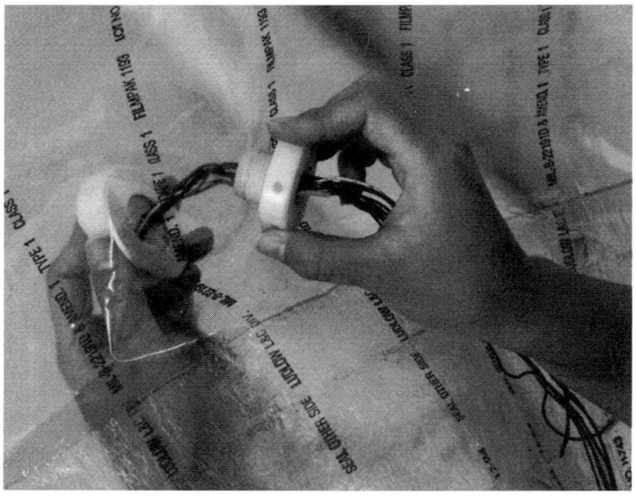

**Figure 4.18**

Hermetically sealed electrical connector being fitted to a film

Repeated tests proved that the 6 $m^3$ Rentokil bubble was suitable for use in nitrogen anoxia even though it required a considerable amount of nitrogen. The oxygen leak rate of 450 ppm (2700 cc) per day was not so great that anoxic conditions of 0.1% oxygen concentration could not easily be maintained over the course of a two-week treatment. However, 86.2 $m^3$ of nitrogen, equivalent to ten T-sized cylinders, were needed, six for the initial flushing and four for the fourteen days of the treatment. (Note that the requirement in this case was a 0.1% oxygen level and the actual level maintained was considerably lower. At least one less cylinder would have been required for a 0.2% oxygen concentration, more than adequate for complete insect eradication.) Liquid nitrogen would be a much better nitrogen source for Bubbles of this size and larger, because a single standard container would have sufficed for the two-week GCI anoxia experiments. A nitrogen generator would have been even more convenient.

A test of an older (1991) 30 $m^3$ Bubble at the Museum of Fine Arts in Houston gave less satisfactory results. Unfortunately, its leak rate proved to be 800 ppm or 24,000 cc of oxygen per day (Daniel et al. 1993). (Greasing or other treatment of the zipper connection between the top and the floor was not done in this first treatment.) Rather than using a continuous flow of nitrogen to cope with this large leakage, a labor-intensive purge of the Bubble to low oxygen levels was done twice each day. As could be predicted, to maintain an oxygen concentration of 0.1% or below for the two weeks of an anoxia treatment required an enormous consumption of nitrogen, 675.5 $m^3$, or more than 78 cylinders of nitrogen. It was only the extremely low cost of nitrogen in Houston (because of its chemical industries) and the great effort exerted by the conservator, Steven Pine, and his aides that made this older Bubble effective. (In retrospect, use of seven tanks of liquid nitrogen would have saved considerable labor, but the GCI research and development work had not been done at this time.) However, the Houston Bubble has been used very successfully many times since the original test. Pine reports that using a small amount of mineral spirits in the groove of the zipper has reduced the nitrogen consumption greatly. Further, more recently produced PVC Bubbles intrinsically may have better seals between top and floor pieces.

Current costs of the Power Plastics PVC Bubbles for 3 $m^3$, 10 $m^3$, and 15 $m^3$ sizes are approximately $1,000, $1,300, and $1,800, respectively, exclusive of shipping costs.

### Modification of Power Plastics Bubbles (used by Rentokil) for anoxia

The experience gained during the test program at the GCI leads to a number of practical recommendations for the effective adaptation of Bubbles to nitrogen anoxia. To begin, the standard equipment necessary for large-scale anoxia in any type of big enclosure must be procured and made ready: a high-volume nitrogen source (liquid nitrogen with regulator, or membrane nitrogen generator), a gas humidification system, oxygen and humidity sensors plus the appropriate monitoring instruments, a vacuum pump and flowmeter, a vacuum cleaner (to act as an exhaust fan in expediting the air flushing of nitrogen from the Bubble), and the items for modifications mentioned above for the 6 $m^3$ Bubble. (Details of these essentials and the operation of a Bubble system are provided in Chapter 5.)

Most portable Bubbles do not come with an internal support, and thus it is strongly advised that a frame be built from PVC pipe and fittings to prevent damage to objects by the relatively heavy plastic film of the Bubble. The plastic zip-lock of the Bubble is its major source of leakage. As previously stated, leakage can be decreased markedly if the zip-lock is lightly greased with laboratory or vacuum grease or even Vaseline. However, great care must be taken not to soil valuable objects—or shoes and tools—with the grease. (Pines's use of small amounts of mineral spirits for this purpose may pose less of a soiling problem.)

During the Rentokil Bubble experimental program, a number of leak spots were found on a large Bubble by checking with a halogen leak detector (see Chapter 2). Unfortunately, it is difficult to fully inspect a Bubble of this size when it is inflated. Some kind of structure, such as that for working from overhead on large museum pieces, must be used so that the entire Bubble can be thoroughly tested for leaks. It is unwise to waste money on enormous quantities of nitrogen in trying to maintain low oxygen levels inside a Bubble that cannot be tested and sealed as a first step. Then a final, definitive leak rate should be determined by the procedure in Appendix I or a full fourteen-day trial treatment carried out to find the nitrogen usage. Because of the normal but acceptable rapid air leak into a Bubble compared to a small bag, the quantitative leak rate could be determined in hours using an adequately sensitive oxygen analyzer. It should take no longer than about two days to see an increase of 1000 ppm in oxygen after the Bubble has been flushed with nitrogen to somewhere near 0.5% oxygen. (A 6 $m^3$ Bubble would require about 36,000 L of nitrogen, 4.5 T cylinders, or a third of a 160 L tank of liquid nitrogen, to reach this level.)

### Leak-proof Connectors to Anoxic Enclosures

The installation of leak-free electrical connectors (for wires to sensors inside the enclosure) and gas connectors (for inlet and exhaust of nitrogen) is a minor facet of the construction of an anoxia enclosure. However, it is not inappropriate here briefly to stress the importance of such an apparent triviality, because treating it too casually in practice was the cause of considerable lost time in early work with anoxic enclosures. Now that the problem has been identified and solved, the detailed discussion of the procedure is provided in the next chapter so that difficulty with leakage around ports or connectors either in large or in small anoxia enclosures will be eliminated.

### Leak Rates of Large Anoxic Enclosures Made from Plastic Film

Large anoxic enclosures made from film have two of the same sources of oxygen leakage as smaller bags: (1) transmission through their walls of plastic film and (2) leaks from imperfect seals and connector holes. A leakage source that small bags do not have is (3) the connection between the base and top portions of large two-piece enclosures such as the GCI tents and PVC Bubbles.

1. All containers made from permeable plastic film laminates allow oxygen to leak into them at a determinable oxygen transmission rate (see Chapter 2; Tables 2.2, 2.3, pp. 21, 22). However, these laboratory results involve very small pieces of film in carefully controlled experiments under special conditions. Frequently in scaling up to practical applications, materials whose physical properties have been measured for small samples in a testing laboratory do not perform as well. Thus it was a principal goal of the GCI's anoxia project to determine the leak rates of both low and higher OTR films in the practical form of 20 L to 1100 L bags, the kind that would often be used by a cultural institution.

Table 4.1 shows the results for an Aclar film whose OTR in air was 10.6 cc/m$^2$/day; Table 4.2 shows the results for a (plausibly) PVDC film with an OTR in air of only 0.06 cc/m$^2$/day. With all the potential errors of making long continuous heat seals and installing connectors in large enclosures, the low OTR PVDC film in an 1100 L bag showed no practical difference from a 20 L bag: 0.28 cc/m$^2$/day versus 0.21 cc/m$^2$/day. In a similar comparison using higher OTR Aclar film, the 1100 L bag had an actual measured OTR of 9.9 cc/m$^2$/day compared to 15.6 cc/m$^2$/day for the 20 L unit. The latter value could have resulted from a proportionately greater contribution from leaks around the three connectors in the smaller bag than in the larger. Overall, the results with these experimental bags show that all sizes of bags can be fabricated so that their leak rates are close to the testing laboratory OTR for Aclar-type "high" permeability barrier films and only three or four times the extremely small OTR of low permeability films. (To put "three or four times" the OTR of low permeability films in perspective: if a film has a laboratory-measured OTR from air of 0.06 cc/m$^2$/day, an 1100 L bag with an actual OTR of "four times" that, or 0.28 cc/m²/day, would still leak less than 2 cc of oxygen per day.)

2. Leaks from imperfect seals and connector holes certainly can be significant, but just as certainly they can be reduced to small quantities by practice in heat sealing and by careful attention to the instructions given in Chapter 5. The kind of concern called for there in making, inserting, and protecting connectors from any accidental movement will result in bags having the excellent leak rates shown in Tables 4.1 and 4.2.

3. A significant amount of oxygen will almost certainly leak into the enclosure through the bottom seal of reusable tents. Obviously, the quantity will depend on the precision or tightness of that temporary seal. Data from the GCI tests using cubical tents made from Filmpak 1193 film showed that the amount of oxygen leaking through the seal depended on its length, the tent perimeter. In a 1 $m^3$ tent in which the seal was constructed as described above, the leak rate was 10 cc/m/day, whereas for a similarly constructed 10 $m^3$ tent, it was 50 cc/m/day. No cause could be found other than that the quality of the seal can be controlled much better in smaller enclosures.

Leakage through the patented zip-lock of a 6 $m^3$ Power Plastics Bubble (labeled "Rentokil") was somewhat greater. Even though this closure was treated with vacuum grease and all other detectable leaks in the Bubble were sealed, the measured leak rate (2723 cc/day) was 540 cc per day higher than the calculated transmission rate of the Bubble's plastic film (2183 cc/day). This was equal to 75 cc of oxygen per meter of seal per day (based on a perimeter of 7.2 m).

**Table 4.1**
Comparison of published and measured OTR values of an Aclar laminate with OTRs calculated from oxygen leak rates of 20 L to 1100 L anoxic bags fabricated from the Aclar laminate

| Method of Testing | OTR in Air, cc/$m^2$/day |
|---|---|
| Calculated from measured OTRs of film in the ASTM standard condition* | 10.6 |
| Measured from $O_2$ leak rate of 20 L bags | 15.6 |
| Measured from $O_2$ leak rate of 230 L bags | 12.8 |
| Measured from $O_2$ leak rate of 450 L bags | 10.8 |
| Measured from $O_2$ leak rate of 1100 L bags | 9.9 |

*OTR in 100% oxygen environment. (The published OTR value is 51 cc/m²/day [Bell Fibre Products, Division of Bell Packaging Corporation, Columbus, GA 31993].)

**Table 4.2**
Comparison of a measured OTR value of a PVDC laminate with OTRs of 20 and 1100 L anoxic bags fabricated from the PVDC laminate

| Method of Testing | OTR in Air, cc/$m^2$/day |
|---|---|
| OTR of film, measured in 100% oxygen and converted to the value in air | 0.6 |
| $O_2$ leak rate of 20 L bags, measured | 0.21 |
| $O_2$ leak rate of 1100 L bags, measured | 0.28 |

*Published OTR in 100% oxygen is 0.1 cc/m²/day, Ludlow Corporation, Homer, LA 71040.

Figure 4.19 is a useful correlation diagram of the oxygen leak rate per day into an enclosure versus the OTR of its plastic film and also shows the consumption of nitrogen over a fourteen-day treatment period to maintain a low oxygen level.

## Rigid Chambers for Anoxia

A museum or other organization that is confronted with continuing insect infestation problems, especially if many objects or large ones are involved, might well consider purchasing a permanent rigid chamber for anoxic treatments. A recommended size is about 3 $m^3$ to 7 $m^3$ because flexible, reusable tents of less than this volume, such as 1 $m^3$, can readily be made with barrier films, as has been described. Furthermore, larger institutions or those with insect-susceptible ethnographic collections might incur greater labor costs over the years with inexpensive reusable tents than with a durable rigid chamber with fully automated operation, despite its high initial capital cost.

There are several manufacturers of environmental or fumigation chambers that can modify their standard models to meet a conservator's requirements for insect eradication with nitrogen. It is preferable to take advantage of their expertise rather than to use a less experienced but more convenient or initially cheaper local metal or rigid plastics fabricator. (In at least one case a "cheap" chamber made locally proved not only to be costly but unusable.) Some requirements that a purchaser should consider are given in the following sections.

In general, the chamber size should be larger than one or two cubic meters because the greater capacity for treatment of either more or bigger objects can readily justify its higher cost. (Even chambers of 3 $m^3$ volume would not allow bulky pieces of furniture to be treated.) A boxlike configuration is preferable to a cylindrical configuration because it does not require installation of a horizontal interior floor and because of

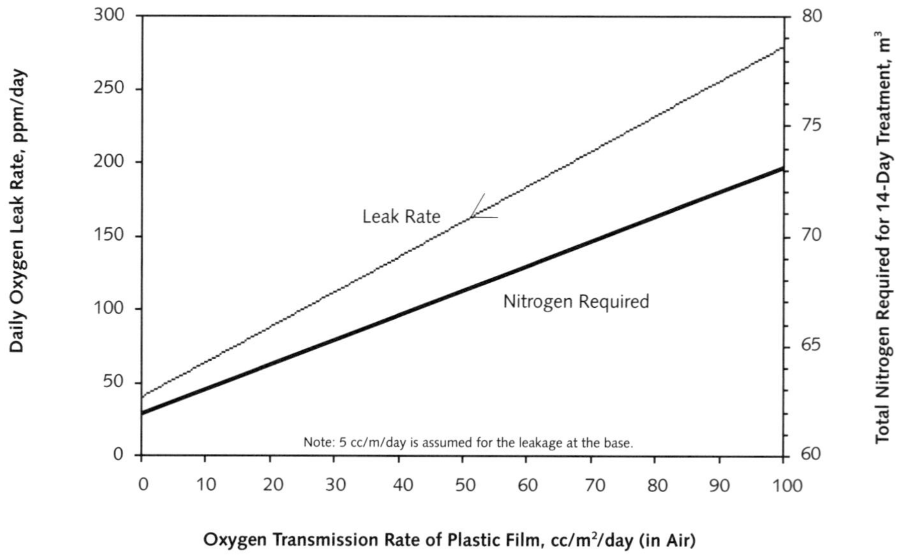


**Figure 4.19**

The nitrogen requirements to flush and to maintain low oxygen levels in a 10 $m^3$ anoxic tent will increase significantly if films with higher oxygen transmission rates are used.

its efficiency in accommodating more objects in a given volume. (Also, the square door of a box is a much more convenient access than the round door of a cylindrical chamber.) Of course, the rigid chamber must be airtight, with an oxygen leak rate of less than 50 ppm per day, and must be equipped with a gauge for low pressure (to 10 inches of water, 2.5 kPa) and with nitrogen entrance and exit valves at the opposite ends.

Other accessories that are essential for rigid chamber operation are sensors and recorders, tubing for the inert gas supply, and a humidification system, like that described above, large enough to bring a rapidly flowing dry inert gas up to the desirable 55% RH or higher. The most convenient nitrogen sources are, in descending order, a nitrogen generator, liquid nitrogen, or multiple cylinders of nitrogen in portable banks. Obviously, the standard safety monitors for oxygen level determination are essential.

It is necessary to equip the chamber with monitoring instruments that will not only monitor but also record the variables of temperature, relative humidity, and oxygen concentration. In addition, the instrumentation should provide control—shut-offs and alarms—at limits that would be harmful to objects or would interfere with the anoxic treatment. This makes automatic operation possible and thereby decreases personnel costs.

### Testing and operation of a rigid chamber

The leak testing of a rigid chamber can be the same as for a flexible enclosure, or an ultrasonic detector may be used (see Chapter 2). Because a rigid chamber has the capability of being slightly pressurized for leak testing to the safe limit recommended by the manufacturer, leaks can be found more readily by the techniques described in Chapter 2. A precise test of the rate of oxygen leakage can be conducted using a trace oxygen analyzer. This leak rate will determine the operational procedure as well as the volume of nitrogen that will be required to maintain a low oxygen level throughout the anoxic treatment. An example is given below. Details of a cycle of treatment are provided in Chapter 5.

A 1 $m^3$ (90 cm wide × 90 cm deep × 120 cm high) near-cubical box of stainless steel was originally purchased by the Los Angeles County Museum of Art (LACMA) from Vacudyne, Inc., for fumigation with ethylene oxide in 1983. It was modified for insect eradication treatment using nitrogen gas from cylinders in 1993. GCI staff collaborated in the testing and initial work using anoxia. An oxygen sensor cell and the temperature and RH sensors were placed inside the chamber through a hermetic electrical connector in the original piping of the system. With the aid of the trace oxygen analyzer and after the chamber had been flushed with nitrogen to a low oxygen level, the daily leak rate of oxygen was found to be 200 ppm per day under 15 cm Hg vacuum, 130 ppm per day at 5 cm vacuum, and 35 ppm per day when a slight positive pressure was kept in the chamber. Therefore, nonvacuum conditions (atmospheric or slightly positive pressure) were clearly important for minimal leakage and normal operation.

In an actual treatment of objects that had been infested, the chamber was first evacuated to about −0.3 barg (−5 psig), and the oxygen content dropped to 17.5%. (This evacuation was a leak check. It need not be used as a normal procedure of anoxia in a rigid chamber. It required undue attention to avoid harm to the humidification system when nitrogen was first admitted to the evacuated chamber. Also, it showed that the nitrogen needed to flush out the air was not greatly decreased by removing 3% of the oxygen under vacuum.) Because this level was maintained for some time, it assured the operator that the door had been adequately tightened and all other openings were securely closed. Nitrogen, humidified to the desirable level of 45% to 55% RH, was slowly admitted to the chamber until it reached atmospheric pressure, at which time a vent valve to the atmosphere was opened and the nitrogen flow increased greatly. It took some twenty-four hours for the chamber to be flushed about eight times (requiring a cylinder of nitrogen, approx. 8 $m^3$) and the oxygen content to drop below 0.1%, that is, below 1000 ppm. The simplest procedure from that point was to stop the input of nitrogen, close all valves to the chamber, and let it stand for ten days until the oxygen content—due to leakage—reached 0.25%, or 2500 ppm. Then the vent and nitrogen inlets were opened, the chamber reflushed to less than 0.2%, the nitrogen access and vent closed, and the chamber left for another four or five days before flushing with air and opening to remove the objects.

The LACMA rigid chamber has been successfully used very many times in the past eight years. More than four anoxia treatments can be carried out with the cylinders of nitrogen in a six-pack.

A larger automated anoxic treatment chamber, 3.3 $m^3$ in the form of a rectangular box 1.1 m wide × 2 m deep × 1.5 m high, was built by a collaboration of Kofloc and Tokkyorika Kogyo, Inc., of Japan at the Tokyo National Research Institute of Cultural Properties in 1998 (Kigawa et al. 1999). Its nitrogen source is a PSA type of nitrogen generator capable of producing more than 2.7 $m^3$ of 99.9% purity nitrogen per hour. Leak testing of the chamber by means of a trace oxygen analyzer showed that it had a leak rate of 50 ppm of oxygen per day. Because of its automated and integrated system of monitors and controls, the system could be operated easily in an automatic purge mode, flushing the chamber with humidified nitrogen whenever the oxygen content rose above a desired level. The system has proved to be capable of maintaining less than a 0.1% oxygen concentration automatically.

## Notes

1. Theoretically, each time an amount of nitrogen equal to the volume of an enclosure is passed into it, the oxygen concentration inside will be lowered by half, e.g., from 20.9% to 10.45% for the first complete volume exchange. (Overall and approximately, it takes eight such volume exchanges to reach 0.1%: 21 → 10.5 → 5.3 → 2.7 → 1.4 → 0.7 → 0.4 → 0.2 → 0.1.) In practice this has been thoroughly proved. However, if the flow of nitrogen is very rapid or the entry and exit points

in the enclosure are not far from each other, thorough mixing of the nitrogen with the residual oxygen is not attained and even fifteen or more volume exchanges of nitrogen could be required to reduce the oxygen to 0.1%.

## References

**Air Products (Air Products and Chemicals, Inc.)**
2001a. *Safetygram-2: Gaseous Nitrogen.* Available at http://www.apci.com/productstewardship/product_safety/safetygrams/index.asp. See Chapter 1, note 1.

2001b. *Safetygram-7. Liquid Nitrogen.* Available at http://www.apci.com/product stewardship/product_safety/safetygrams/index.asp

2001c. *PRISM® Nitrogen PSA Systems.* www.apci.com/gases/onsitesupply/psa.html, 1–3.

2001d. *PRISM® Nitrogen HPN Systems.* www.apci.com/gases/onsitesupply/hpn.html, 1–3.

2001e. *PRISM® Nitrogen Membrane Systems Cabinet Series.* www.apci.com/gases/onsitesupply/cabinet.html, 1–4.

2001f. *PRISM® Nitrogen Membrane Systems High-Capacity Series.* www.apci.com/gases/onsitesupply/highcap.html, 1–4.

**Daniel, V., S. Maekawa, F. Preusser, and G. Hanlon**
1993. Nitrogen fumigation: A viable alternative. In *ICOM Committee for Conservation 10th Triennial Meeting, Washington, DC, 22–27 August 1993*, 863–67. Marina del Rey, Calif.: ICOM Committee for Conservation.

**Elert, K., and S. Maekawa**
1997. Rentokil Bubble in nitrogen anoxia treatment of museum pests. *Studies in Conservation* 42:247–52.

**Kigawa, R., K. Yamano, S. Miura, and S. Maekawa**
1999. An automated anoxic treatment chamber for cultural objects—A novel system with a nitrogen generator. *Science for Conservation* 38:1–9. Tokyo: Tokyo National Research Institute of Cultural Properties.

**Kirk, R. E., D. F. Othmer, J. I. Kroschwitz, and M. Howe-Grant**
1996. *Encyclopedia of Chemical Technology*, vol. 17. 4th ed. New York: Wiley.

**Maekawa, S., and K. Elert**
1996. Large-scale disinfestation of museum objects using nitrogen anoxia. In *ICOM Committee for Conservation 11th Triennial Meeting, Edinburgh, Scotland, 1–6 September 1996, Preprints*, vol.1, ed. J. Bridgland, 48–53. London: James and James.

Chapter 5

# Protocols for Insect Eradication in Nitrogen (Anoxia)

This chapter describes four anoxia protocols that can be used to exterminate insects attacking museum objects of various sizes. Although the instructions for each protocol are self-explanatory, it is highly recommended that the preceding chapters be read first to understand the principles underlying each treatment option. This chapter is divided into two parts:

- *Choosing a Protocol:* brief description of each protocol and a table to help choose the appropriate protocol.
- *Anoxia Protocols:* step-by-step instructions for each protocol, including a troubleshooting section and a list of required supplies and equipment.

Appendix I, Protocol Tools, explains the basic tests, techniques, and calculations that are referred to in each protocol. The reader is advised to read Appendix I before attempting treatment.

Appendix IV provides contact information, including Web sites, for the manufacturers and suppliers of anoxia materials and equipment referred to in this chapter.

Appendix V provides outlines for treatment protocols A, B, and C. We recommend that readers use them in treatments as a checklist.

## Choosing a Protocol

The four anoxia protocols, which are briefly introduced here, are tailored to the number and size of infested objects and to their anticipated storage time under anoxic conditions. Table 5.1 shows how to select the appropriate protocol.

**Protocol A** is the most basic and least expensive procedure. It is used for small objects treated in small to medium-size bags made from sheets of plastic film or in premade pouches. These bags and pouches range in volume from about 0.001 to 1 $m^3$ (1–1000 L). A semiautomatic impulse heat sealer makes it possible to quickly fabricate a large number of pouches at once. If treated objects are to be stored for a year or more under anoxic conditions, treatment bags should be made of ESCAL™, a special moisture- and oxygen-barrier film.

| Quantity of Objects | Examples | Protocol to Use |
|---|---|---|
| Fewer than 100 small objects | Newly acquired or donated books, dolls, and clothes; small musical instruments such as violins | Protocol A |
| A few to several medium-size objects | Panel paintings; flat textiles such as folded cloths and flags | Protocol A |
| Fewer than 100 small objects requiring long-term storage | Archaeological metals such as coins, bells, knives; rubberized garments such as a spacesuit or raincoat | Protocol A, using ESCAL™ film for storage |
| Hundreds of small objects requiring long-term storage | Same as above, plus mummified human remains | Protocol A, using ESCAL™ film for storage and an impulse sealer for mass bag production |
| Hundreds of small objects | Herbarium and natural history collections; large library and archive collections | Protocols B and D, with shelving |
| A dozen or more medium-size objects; a few large pieces | Furniture pieces in various sizes; large textile collections such as carpets; large ethnographic collections; large musical instruments such as pianos | Protocols B and D |

*Notes:*
Protocol A—Anoxia in small to medium-size bags or pouches
Protocol B—Anoxia in large, user-made tents
Protocol C—Anoxia in large commercial enclosures (Portable Bubbles)
Protocol D—Anoxia in large rigid chambers

**Table 5.1**
Choosing a protocol

*Nitrogen purging:* In this variation on Protocol A, the anoxia bag is flushed with nitrogen to accelerate the treatment process and reduce the amount of oxygen absorber required.

**Protocol B** uses large, user-made plastic-film tents typically 2 $m^3$ to 10 $m^3$ in size.

**Protocol C** uses large commercial enclosures known by the trade name Portable Bubbles, approximately 3 $m^3$ to 30 $m^3$ in size.

**Protocol D** uses large, rigid chambers, approximately 3 $m^3$ to 10 $m^3$, such as the commercial units previously used by museums to fumigate objects with ethylene oxide or similar toxic substances.

*Note:* Protocols B through D can be used with shelving to allow proper circulation of the anoxic atmosphere when treating multiple objects.

## Protocol A: Anoxia in Small to Medium-sized Bags or Pouches

Protocol A is the most basic and least expensive anoxia procedure. It is used for small to medium-size treatment bags up to 1 $m^3$ in volume. Infested objects are sealed in the plastic bag or pouch, along with an oxygen absorber, an oxygen monitor, a humidity buffer, and a humidity monitor. Treatment typically takes fourteen days.

Treatment bags are easily made with a variety of oxygen-barrier films (see Chapter 2, pp. 22, 24–27). One product, ESCAL™, is a superior moisture- and oxygen-barrier film. Bags and pouches made from this product are ideal for storing treated objects for up to ten years.[1]

Premade pouches and bags are also available from conservation suppliers (see Appendix IV), and these are especially useful when a large number of bags are needed. Premade bags have internal dimensions ranging from 16 cm × 24 cm to 76 cm × 84 cm. Other sizes can be made to order but at greater cost.

Individuals using Protocol A for the first time can achieve successful results. However, one or more trial runs with dummy objects are recommended. This will improve the user's proficiency in making well-sealed treatment bags and testing them for leaks without harming valuable museum objects.

The basic technique of fabricating a treatment bag is described below. This is followed by a discussion of nitrogen purging to reduce the amount of oxygen absorber needed for treatment under certain circumstances. A list of materials and equipment needed to fabricate and use anoxic treatment bags is provided at the end of this section.

**Step 1 (build protective frame):** When the object to be treated is fragile and should not be weighed down, even by a relatively light plastic film, build a simple sturdy frame to protect it. This frame need not be elaborate: a box open at the top and bottom and built slightly larger than the object is usually adequate (Fig. 5.1). Several frames can be used in a single anoxia bag when treating a number of objects at one time. A wire frame, bent from 9 gauge (B&S) wire, is also an easily made protector.

**Step 2 (make a pattern):** Use an inexpensive plastic film, such as polyethylene, to create a pattern for the exact size of the oxygen-barrier film that will be used to form the treatment bag. Making a pattern will avoid waste, especially if costly film will be used for treatment. The pattern will be sized using a substitute for the actual object (or objects) to be treated. The substitute object can be the protective frame (if needed), a cardboard mock-up, or a dummy that is the same size and shape.

The piece of plastic to be used for the pattern should be about 5 times the length and 4 times the width of the dummy object. If the dummy is more than about 20 cm (8 in.) high, start with a piece of film that is 6 to 7 times longer and 5 to 6 times wider. One-half of this length will form the underside of the bag. Center the dummy object on this half and place in front of or beside it one or more empty plastic

**Figure 5.1, below**
A frame protects fragile objects from being damaged by the plastic film.

**Figure 5.2, below right**
The plastic film is folded loosely over the object to create a pouch (bag).

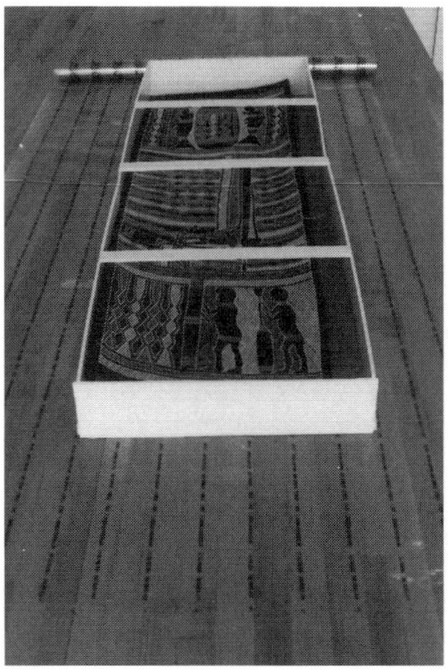

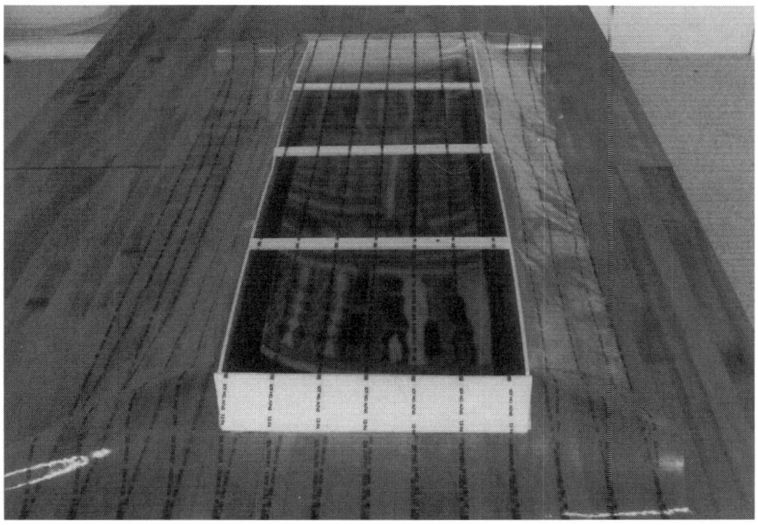

trays large enough to hold the RH-conditioned silica gel that will take up the water released by the oxygen absorber during treatment. (See the Technical Notes on oxygen absorbers/silica gel at the end of this section.) The need for a humidity buffer is discussed in Chapter 3, pages 53–56. RH monitors are discussed in Chapter 2, pages 36–37.

Fold the film in half over the dummy and silica gel trays so there is an easy drape on all sides, as in Figure 5.2. The plastic film can now be trimmed to make a pattern. Leave at least 8 cm to 10 cm of plastic lying perfectly flat along the two open side seams and along the fold (this is particularly important for objects whose height is significantly greater than their width and length; the film has to be wide enough to lie flat at the edges and allow wrinkle-free seams). Allow 50 cm to 75 cm of flat film for the seam opposite the fold. This will be the front of the bag and the last edge to be sealed. This excess front-edge film permits reuse of the actual bag after the 5 cm seal is cut off to open the bag after treatment. Before cutting, make sure that the protective frame or dummy object does not pull or stretch the plastic.

**Step 3 (cut the oxygen-barrier film):** Lay the oxygen-barrier film to be used for the treatment bag on a flat surface with its heat-sealable side facing upward. If the film will be cut with a razor blade or knife rather than scissors, place cardboard or other protection under it. Next place the pattern created in Step 2 over the barrier film and with a permanent marker, mark the corners of the pattern on the film. If making a large treatment bag, it helps to use a 4 ft carpenter's metal rule to connect the corner marks and then draw in the cut lines with the marker. Cut the oxygen-barrier film.

**Opaque film:** If aluminized (opaque) film is used, a window should be created in the bag so the object and oxygen monitor can be viewed during treatment (Fig. 5.3). The window opening should be 10 or 20 cm square and cut into the opaque film such that once the film is folded and sealed into a bag, the window will be located on top in a convenient viewing area but not close to any edge to avoid interference with the final sealing of the bag (Fig. 5.4). The window cutout is covered with a square of clear oxygen-barrier film that is 18 or 28 cm on a side, depending on the size of the opening made. Heat seal the sealable side of the clear window to the corresponding side of the opaque film, which should be the side facing up on the work surface.

**Figure 5.3, below**
A window allows monitoring the oxygen indicator and the condition of objects during anoxic treatment.

**Figure 5.4, below right**
Transparent film is sealed over a cut-out in an opaque bag to form a window.

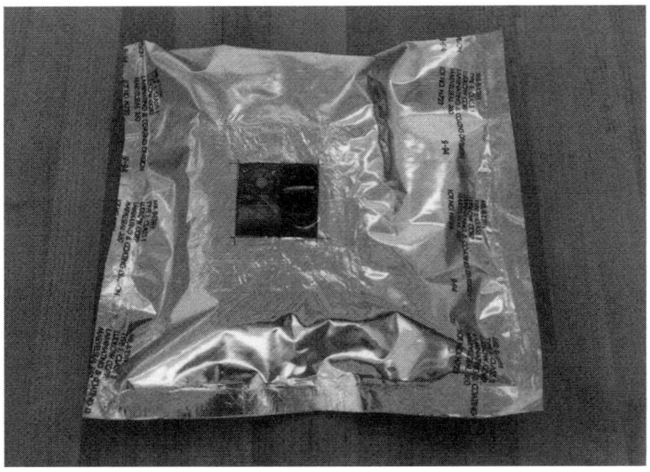

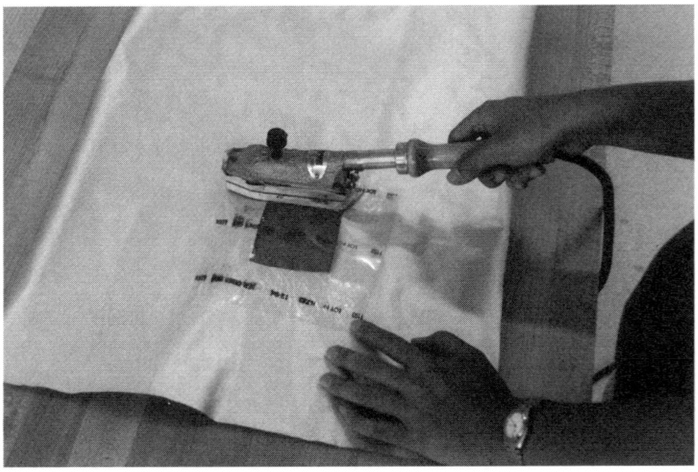

**Step 4 (assemble and leak test the treatment bag):** Before folding and sealing the oxygen-barrier film into a treatment bag, make sure any sharp edges on the object(s) or protective frame are covered with cardboard guards. If needed to protect the film, have ready a protective base of cardboard or soft open-pore foam to be placed under the object. (Do not use bubble wrap because it will slowly release oxygen from its trapped air.) Now proceed as follows:

(a) Check again that the heat-sealable side of the film is facing up. Then place the object (or a dummy if this is a trial run) directly on the plastic film, with a protective frame and base if needed, positioning the object so that it will be roughly in the center of the finished bag. It should be slightly closer to what will be the front edge of the bag to avoid straining or pulling on the object at the fold when the film is folded in half.

(b) Fold the plastic film over the object, lining up the edges that will form the front of the bag. Using a tacking iron or bar sealer,[2] heat seal together the two open sides and part of the front (Fig. 5.5), leaving an opening in the front large enough to slide in the trays of silica gel and a humidity monitor. (Use empty trays and no monitor if this is a trial run.)

(c) Insert through the opening two colored oxygen monitors, such as Ageless-Eye®, fixing them to the floor of the bag with tape and making sure they are visible through the window in opaque film. An alternative is to use a small battery-operated oxygen monitor in the bag, such as shown in Figures 2.7, 2.8, and 2.11 (see pp. 33–35). Or plan to take daily or twice-daily samples of the bag's atmosphere with a head-space oxygen analyzer (see Chapter 2, p. 35, and Fig. 2.9).

(d) Disperse (but do not heap) the required number of Ageless® Z oxygen absorber packets around the object being treated, making sure that the packets do not touch the object (Fig. 5.6). (See the Technical Notes on oxygen absorbers at the end of this section.) The packets can be

**Figure 5.5, below**
Initial heat sealing of the bag

**Figure 5.6, below right**
Oxygen absorber packets and humidity buffering material (RH-adjusted silica gel) are placed in the bag after most of the heat seals have been completed.

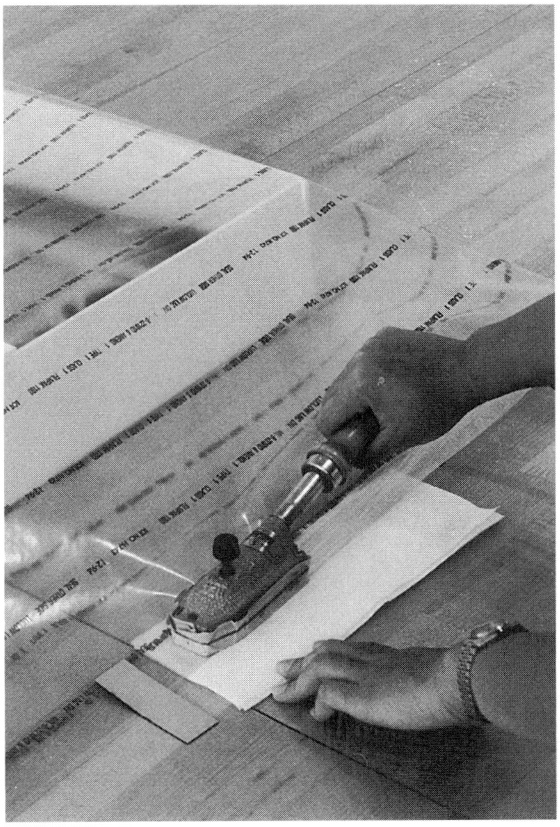

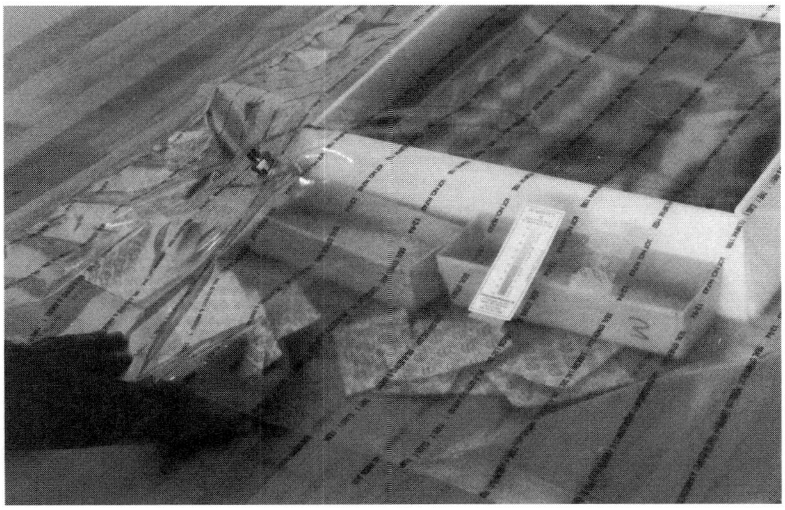

stapled together to form a long strip that is easy to insert into the bag. Oxygen absorbers are described in Chapter 3, pages 43–57. Optional: the amount of oxygen absorber required can be greatly reduced by purging the treatment bag with nitrogen after it has been filled and leak tested, as described in the next section.

(e) Heat seal the remainder of the front edge except for a 1 cm to 2 cm opening in acorner. This will be used to insert halogen gas for leak testing.
(f) Leak test the bag (see Appendix I).
(g) Optional: Purge bag with nitrogen to reduce amount of oxygen absorber required. See next section for details about nitrogen purging.
(h) Let the bag sit undisturbed for fourteen days at temperature of at least 25°C.

***Caution:*** Some conservators slightly evacuate the anoxia bag before beginning treatment to remove a small amount of air to save several packets of oxygen absorber, as shown in Figure 5.7. *This is not advised.* The savings are not great. More important, the objects being treated may be damaged. During treatment, when oxygen is being absorbed, the volume of the bag is reduced by one-fifth. If the bag has already been evacuated slightly at the start, the further reduction in volume during treatment may result in a virtual "shrink wrap" of a valued object.

**Step 5 (monitor oxygen level):** Throughout the fourteen-day treatment, check the oxygen monitor in the bag at least daily (or sample the bag's atmosphere with a head-space oxygen analyzer) to ensure that the oxygen content stays at or below 0.3% after stabilizing. If an Ageless-Eye® monitor is used, it should change from violet to pink within two to two and a half days. If it does not, follow the instructions in the Troubleshooting section at the end of Protocol A.

If the Ageless-Eye® changes to pink in two days after the start of the treatment but then returns to violet at any time, this means that a very large quantity of oxygen has leaked into the bag. If this happens,

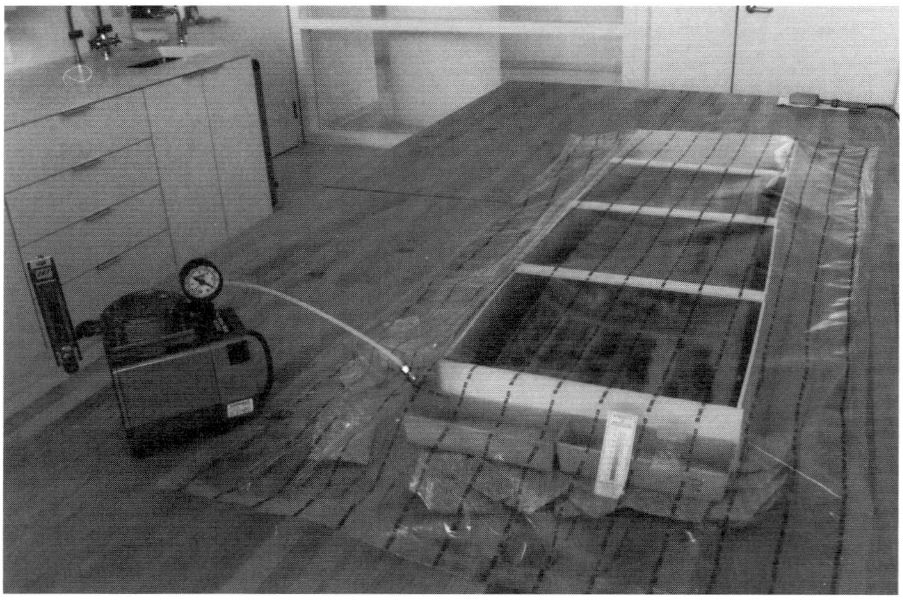

**Figure 5.7**
Evacuation of the bag using a vacuum pump (not recommended)

repeat the treatment with a new bag and carefully inspect the used bag to find out why such a significant leak was not detected during leak testing at the start of the process.

**Step 6 (end treatment):** At the end of fourteen days, open the treatment bag. Cut the front seam off as close as possible to the seal to minimize waste so the bag can be reused for another treatment. Remove the disinfested object and accessories from the bag.

Place the oxygen absorber packets in a loosely closed polyethylene bag for a day so that any unused content will slowly react with air, then discard them. Because oxygen absorbers generate heat as they react with oxygen, make sure that the packets are kept separate and not piled together to dissipate the heat buildup.

Aluminized film develops pinholes easily, and bags made from this opaque film should be inspected carefully before reusing. Cover the transparent window with any opaque material and hold the used bag up to sunlight. Look through the open end for any spots of light that would indicate pinholes.

*Technical Notes on Oxygen Absorbers/Silica Gel*

1. The calculation used to determine how much oxygen absorber is needed is found in Appendix I.

   For a sealed pouch that is 30 cm × 25 cm, the standard equation of length (cm) × width (cm) × height of 8 (cm) × 0.00021 yields 30 × 25 × 8 × 0.00021 = 1.3 packets of Ageless® oxygen absorber of 1000 cc capacity. In this case, two 1000 cc packets of oxygen absorber would be sufficient.

   If the object is to be stored in the sealed bag for a year, the oxygen leak rate of the film must be taken into consideration. The leak rate of some oxygen-barrier films is about 4 cc/m$^2$/day. Therefore, in the above example, the two surfaces of the bag have a total area of 2 × (25 cm × 30 cm) = 1500 cm$^2$, or 0.15 m$^2$. Thus 219 cc of oxygen would be expected to leak into the pouch over a 365-day period, using the calculation 365 days × 4 cc/m$^2$/day × 0.15 m$^2$. The total amount of oxygen originally in the pouch (1300 cc) plus the quantity that leaked in over one year (219 cc) yields 1519 cc of oxygen. This is well within the actual capacity of two Ageless® Z-1000 or Atco™ LH 1000 packets.

2. The calculation used to determine the correct amount of silica gel is found in Appendix I.

   The moisture released by an Ageless® Z-1000 packet over time at 50% RH is 0.4 g. Fifty grams of silica gel already conditioned to a 50% RH atmosphere will react with 1 g of water before it changes its equilibrium value to 55% RH. Therefore, this amount of silica gel will strongly buffer the RH inside the pouch at near 50% RH. (If the water transmission rate for the film used to make the above bag is 0.5 to 5 g/m$^2$/day, this would mean a maximum rate of 0.075 g to 0.75 g of water per day in or out of the bag. This rate could be buffered by the silica gel before and as it reaches an equilibrium value with the changing RH inside the bag.)

*Note:* If an object such as a massive piece of furniture already contains a lot of moisture, the amount of moisture released by oxygen absorbers is not significant, and silica gel is not required to counter the resulting humidity change to 75% RH. Among objects that are not sensitive to the increased humidity are many textiles, lacquer ware, and some less fragile furniture.

### Nitrogen purging: Reducing the amount of oxygen absorber

Initially purging the treatment bag with nitrogen will speed up the time it takes the atmosphere to drop below 0.1% oxygen and will assure that it is maintained for the standard fourteen-day treatment period. Purging dilutes the oxygen in the bag. Purging with a quantity of nitrogen equal to the bag's volume reduces the amount of oxygen absorber needed by one-half. Nitrogen purging with 8 times the volume of a well-sealed pouch will reduce the required amount of oxygen absorber to one-tenth the nominal quantity required for nonflushed bags (the maximum reduction possible); this is enough to absorb any oxygen that seeps out of the objects being treated as well as oxygen that might seep into the bag (although oxygen seepage into small to medium-size bags that are well constructed is insignificant).

In general, nitrogen purging is advisable for treatment bags between 100 and 1000 L in volume. Purging is cost-effective with even smaller bags, but only if there are many bags, in order to offset the cost of setting up the purging system. A steady flow of nitrogen rather than oxygen absorber is used to create an anoxic environment in bags 1000 L (1 $m^3$) and larger. This is described in the protocols for large-scale treatment systems.

An initial nitrogen purge can generate appreciable cost savings, especially if many treatments are planned. For example, in 2002 the purging costs associated with 10 treatments using a 100 L bag amount to about $30 for 8 $m^3$ of nitrogen and about $20 for 20 packets of oxygen absorber, for a total savings of about $50. Without the nitrogen purge, 10 treatment runs would use 210 packets of oxygen absorber, costing about $210.

However, the initial cost of the nitrogen equipment is large—perhaps five times the savings in a single run. A safe cart for cylinder transport and cylinder supports costs $200, the nitrogen gas regulator and flowmeters $300, the vacuum pump about $300, and the nitrogen humidification apparatus around $250. This total cost is approximately the same as some suppliers charge for 1,050 oxygen absorbers of 1000 cc capacity. This amount of oxygen absorber would be enough for 50 anoxia treatments in a 100 L bag.

See Chapter 4 for background information. The construction of a nitrogen humidification system is described in Appendix I.

*Note:* The preceding comments about costs assume that the objects being treated are sensitive to shock from the rapid RH change caused by the dry nitrogen. The instructions for nitrogen purging below also assume this kind of sensitivity, and thus they specify the use of humidified nitrogen.

If the objects are not endangered by a rapid—but brief—RH change, then exposure to low humidity for the 2 hours or so it takes to purge, for example, a 100 L bag's volume 8 times is not a hazard to them. In that case you can purge with nitrogen that is *not* humidified. After purging, moisture will be released by the object and equilibrate the RH in the bag.

### Nitrogen purging: Physical setup and process

Nitrogen purging is done (1) after the treatment bag is filled with the infested objects, monitors, buffered silica gel, and the reduced quantity of oxygen absorber and (2) after the filled bag has been tested for leaks. If the treatment requires humidified nitrogen, make sure the humidification system is on-line with the cylinder of dry nitrogen.

*Attaching humidified nitrogen source to treatment bag*

Through the 1 cm to 2 cm opening left in the sealed bag for leak testing (see Protocol A, Step 4e), insert about 5 cm of the nylon tubing that comes from the inlet flowmeter attached to the nitrogen line of the humidification system (see Fig. 4.10, p. 68). (If no humidification is necessary, simply insert the tube from the nitrogen cylinder.) Tape the bag closed around the tubing, preferably using masking tape to make removal easier after purging. This is now the nitrogen inlet to the bag. Make sure that the end of the tubing inside the bag does not point directly at an object (aim for the plastic box containing the buffered silica gel to disperse the nitrogen). Cut a diagonal 2 cm to 4 cm exit hole across the corner in the bag most distant from the front opening.

*Setup for bags with Swagelok connectors*

Airtight Swagelok connectors should be installed in large bags to prevent possible leakage at the inlet and outlet points. These connectors are especially handy if you expect to perform more than one purge during a treatment, for example, when treating a large wood object that may take more than 24 hours to diffuse its stored oxygen. (See "Leak-proof Connectors in Anoxic Enclosure Walls" in Appendix I.) Connect the nylon tubing from the inlet flowmeter of the humidification system to the Swagelok-fitted inlet to the bag. Next attach a flowmeter and a vacuum pump to the outlet connector. The pump's control valve should be in a loop circuit arrangement to adjust the amount of gas exhausted from the bag. This arrangement is shown in Figure 4.1 (p. 59), which is a schematic for a large anoxia tent, but the same arrangement of flowmeter and vacuum pump at the outlet is used for bags as well. Turn the dry and wet flow valves on the humidification unit to the settings that produce a 50% RH (or other desired RH for the objects being treated). Establishing these settings is explained in Steps 7 and 8 for the humidification unit description in Appendix I.

*Setting the nitrogen flow rate*

The suggested nitrogen flow rate into the treatment bag for both dry and humidified nitrogen is one-tenth the volume of the bag per minute. This is very approximate but useful for anoxic enclosures about 300 L or less in volume. At this rate, a 100 L bag, for example, would be purged once by nitrogen in 10 minutes and 8 times in about 1.5 hours. (The flow rate for purging with dry nitrogen can be much more rapid to achieve a low-oxygen environment as quickly as possible so that the moisture of the treated object can condition the RH in the bag.)

The nitrogen exit rate must be carefully monitored to equal the entry rate so the bag does not become overinflated and stress a seam, causing leaks. For a bag without a Swagelok connector, the exit rate is readily controlled by simply cutting a larger exit hole or heat sealing the existing hole to a smaller size. If there is a Swagelok connector on the bag, the exit valve can be adjusted. However, if an operator cannot be present to adjust the exit flow throughout the purging process, for example, when purging a large treatment bag with a relatively low flow rate as described next, then a vacuum pump and flowmeter must be attached to the exit to balance the flows.

The one-tenth-the-volume-per-minute flow rate is problematic for an anoxic treatment enclosure as large as, for example, 600 L. In this case, the flow rate would have to be 60 L per minute, and this would require a significantly larger humidification unit than the one detailed in this book or two units in parallel to produce a 50% RH nitrogen output. However, a flow rate of 30 L per minute from a single humidifier is achievable and can result in a 50% RH nitrogen output. Using that more conservative rate, purging a 600 L bag 8 times would take slightly less than three hours. Thus for these large bags, 25 L to 30 L per minute is the preferred rate of nitrogen purging. If you are using a vacuum pump with a control valve and flowmeter to balance the flow of humidified nitrogen, slowly open the inlet valve to the humidification apparatus and let nitrogen begin to flow. Gradually establish a rate of flow per minute that is one-tenth the size of the bag and is balanced with the outflow from the bag. For later reference, make note of the time when the first flow balance has been established and at what flow rate.

If you are not using humidified nitrogen, slowly increase the rate of nitrogen flow from the nitrogen cylinder until the rate shown on the inlet flowmeter is one-tenth of the bag's volume per minute. Be sure that you do not overinflate the bag. If the bag begins to be become distended, cut a larger exit hole in the bag. (If you have a bag with a Swagelok connector as the nitrogen exit and are *not* using a vacuum pump, cut an additional exit for nitrogen in a corner of the bag on the end opposite the nitrogen entry.) For later reference, make note of the time when you attain the one-tenth volume rate.

***Caution:*** Observe the bag frequently during purging to make sure it does not overinflate. The bag can easily become too taut in a matter of minutes, causing a seam to leak.

### *Terminating purging*

The following instructions explain the termination process under two different setup conditions.

1. **If you have an electronic oxygen monitor in the treatment bag:** Terminate purging when the monitor reads an oxygen concentration of about 0.1%. Shut down the nitrogen flow in one of the two following ways:
   (a) If you are not using humidified nitrogen, untape the nitrogen tubing leading into the bag and remove the tubing *while the nitrogen is still flowing*. Immediately heat seal the entry hole, turn off the nitrogen, then quickly heat seal the exit hole. Make note of the elapsed time for the purging and calculate how much nitrogen was used for future reference by multiplying the flow rate by the purging time.
   (b) If you are using humidified nitrogen, as quickly as possible do the following: (i) close the inlet valve from the nitrogen cylinder to the humidification unit; (ii) shut the nitrogen inlet valve to the treatment bag; (iii) turn off the vacuum pump; (iv) close the outlet valve between the exit and the pump. Relieve any pressure in the humidification unit by opening its relief valve. Make note of the elapsed purge time and multiply it by the flowmeter rate to calculate the amount of nitrogen used.

   After terminating purging, begin the fourteen-day treatment, starting with Step 5 in Protocol A.

2. **If you do not have an electronic oxygen monitor in the treatment bag:** Terminate purging when the balanced inlet and outlet flowmeter reading (volume per minute), multiplied by the elapsed time of the nitrogen purge, equals 8 times the bag's volume. Shut down the nitrogen flow in one of the two ways described above.

   After terminating purging, begin the fourteen-day treatment, starting with Step 5 in Protocol A.

## Troubleshooting Protocol A

Small-scale anoxia treatments using plastic bags and oxygen absorber are simple and reliable. Serious problems rarely develop with Protocol A. This troubleshooting section addresses some of the minor difficulties that may be encountered.

*Symptom:*

- The oxygen indicator tablet does not change to pink within two to four days: See Causes 1–4.
- The electronic oxygen monitor does not show oxygen below 0.2% after two days: See Causes 2–4.

*Possible Causes:*

1. The oxygen indicator tablet is defective.
2. There are major leaks in the bag due to imperfect seals or pinholes.
3. Not enough oxygen absorber was placed in the bag.
4. The oxygen absorber is exhausted and no longer reacting with oxygen.

*Solutions:*

1. The oxygen indicator tablet is defective. The quickest, most certain way to determine if the oxygen indicator tablet (such as Ageless-Eye®) is defective is to measure the oxygen concentration inside the treatment bag with an external oxygen analyzer. The preferred instrument is a head-space analyzer (see "Oxygen Monitors and Analyzers" in Chapter 2). With this unit, a syringe needle is simply inserted into the bag and a sample of the atmosphere is drawn into the analyzer.

   If the oxygen content is greater than 0.3%, the oxygen indicator tablet is accurate, and one of the remaining solutions should be consulted. Once the source of the problem has been identified and addressed, the anoxia treatment should be restarted with a new bag, fresh materials, and especially careful sealing of the bag.

   However, if the oxygen in the bag is within the useful anoxia values of 0.1% to 0.3%, the oxygen indicator tablet is clearly defective. Remove the syringe, seal the puncture hole with aluminum tape, and continue the anoxia procedure. Check the oxygen content of the bag every day with the head-space analyzer to assure that adequately low oxygen levels are maintained.

   An indicator tablet may appear defective because the temperature or humidity inside the bag is less than optimal, causing an unduly slow color change to pink. This can happen when the humidity is less than 30% RH or the temperature is below 20°C. This low temperature is particularly problematic because it slows the oxygen absorber's reaction time and a complete kill of insects is not achieved within fourteen days.

   If humidity and temperature are optimal, then the indicator tablet may have failed because it is too old or was stored under unfavorable conditions before use. In this case, the entire stock of tablets from which the defective table was taken must be tested as follows.

   Remove all tablets from their storage pouch and expose them to air until they just turn violet. Then replace them in the pouch with a fresh packet of Ageless® Z-1000 oxygen absorber, and completely seal the pouch. The oxygen indicators that do not turn pink within twenty-four hours are defective. This suggests that the tablet being used in the current anoxia treatment is also defective.

   However, if all tablets turn pink, then the tablet currently being used in the anoxia bag is correctly reporting a high level of oxygen, and one of the remaining solutions should be consulted to determine the source of this excess oxygen. Under these circumstances, it is best to repeat the anoxia treatment with a new bag and with fresh Ageless® Z-1000 and two fresh Ageless-Eye® indicators. Seal the bag carefully with a double rather than a single seam. A bar sealer is preferable because it heats the plastic from both sides, resulting in a better seal than can be obtained with a less expensive tacking iron.

2. There are major leaks in the bag due to imperfect seals or pinholes. Check the bag very carefully for imperfections in the seals or for puncture holes that may have been caused when the infested objects were placed in the bag. Pinholes can be covered for the duration of an

anoxia treatment with self-adhesive aluminum tape. It is often easy to correct imperfections in a seal by making an additional seam around the defect with a tacking iron. If leak points cannot be located visually, they are usually detectable with a halogen leak detector. See Appendix I for instructions on locating and fixing leaks.

3. Not enough oxygen absorber was placed in the bag. Recheck the calculation for the amount of oxygen absorber required for the size of the treatment bag (see Appendix I). Then verify that the correct number of oxygen absorber packets are in the bag.

4. The oxygen absorber is exhausted and no longer reacting with oxygen. The reactivity of the oxygen absorber packets can be checked indirectly without opening the treatment bag by feeling the packets through the bag to see if the contents are clumped or lumpy. This usually indicates that the powdered iron oxide has become exhausted from reacting with too much oxygen. If this is the case, restart the treatment, preferably with a new treatment bag, because there must have been a very large leak in the bag that failed. Before adding new new oxygen absorber packets to the bag, check that their contents are loose and not clumped. A more precise and preferable test of the absorber's efficacy can be done using a sample packet from a batch of oxygen absorbers with the capacity to absorb all the oxygen in a small bag. Seal the sample packet in the bag with an oxygen monitor or, alternatively, check the bag's atmosphere daily with a head-space oxygen analyzer. The results will indicate the reactivity of the oxygen absorber and whether the rest of the batch is good.

### Materials and equipment for fabrication and operation of anoxic bags

See Appendix IV for contact information for manufacturers and suppliers.

**For bag fabrication:**

- Oxygen-barrier film (see Chapter 2, pp. 19–27, and Table 2.3, p. 22)
- Inexpensive polyethylene film (for making practice treatment bags)
- Tacking iron, Teflon coated, temperature regulated: local photographic store, conservation suppliers, McMaster-Carr 1983T2
- Bar sealer, temperature regulated, or impulse bar sealer: Conservation Support, Conservation by Design, IMPAK, Metode, McMaster-Carr 2064T16
- Large ruler; carpenter's 4 ft rule; permanent marker; masking tape, narrow, wide; scissors; sharp knife or razor-blade knife; paper towels; small roll of 18- to 20-gauge wire.

**For leak testing of bag:**

- Halogen leak detector kit (sensor-detector-monitor, halogen gas)
- Model 5750 Refrigerant leak detector, TIF Instruments or Model FGT-202 Refrigerant leak detector, American Gas

- Can of tracer halogen gas (local refrigerant or computer supplier)

**For anoxia process:**
- Ageless® Z packets of oxygen absorber or equivalent: conservation suppliers
- Ageless-Eye® oxygen indicators: conservation suppliers, Mitsubishi
- Electronic oxygen monitor:
  - (a) small or sensor plus external monitor: Teledyne 320B/RC (with remote sensor) or Teledyne 3190 (trace analyzer with remote sensor)
  - (b) head-space analyzer, hand-held instrument: PBI-Dansensor Checkpoint; or instrument with printer: PBI-Dansensor Checkmate
- Bead silica gel: Conservation Support
- Plastic trays for silica gel
- Hygrometer (RH monitor) for anoxia enclosure and for calibrating humidification system: Cole-Parmer U-37100-15, Conservation Support

**For nitrogen source:**
- Nitrogen gas cylinder with safety transport cart and support: Air Liquide, Air Products, Matheson TriGas, Praxair
- Nitrogen two-stage regulator, with gauges: Cole-Parmer, U-98202-23
- Nylon tubing, 1/4 in.: Cole-Parmer U-06489-22 (specific for use in assembling humidification system and for introducing nitrogen into treatment bag/enclosure)
- Copper tubing: Swagelok CU-T4-S-050-50 (1/4 in., 50 ft) with fittings to connect regulator to humidification unit

**For humidification system:**
- See Appendix IV for supply list.

**For nitrogen purging:**
- Flowmeters, 4–40 L/min of nitrogen: 2 Cole-Parmer, U-03216-38
- Vacuum pump, diaphragm, 31 L/min capacity: Cole-Parmer, U-07061-2

**For connectors** (between humidification system and treatment bag/enclosure, from enclosure to vacuum pump, and to flowmeters):
- 2 Swagelok O-ring straight thread male connectors, brass, B-400-1-OR (1/4 in.) with 1/2 in. hole washers, 7/16 in. × 20 nuts)
- 2 Whitey ball valves, brass, B-42S4 (1/4 in.)
- 1 Whitey needle valve, brass, B-1RS4 (1/4 in.)
- 5 port connectors, brass, B-4012-PC (1/4 in.)
- 2 union tees, brass, B-400-3 (1/4 in.)
- 4 female elbows (for flowmeters), B-400-8-2 (1/8 in. NPT 1/4 in. tubing)

- 2 male connectors, B-400-1-4 (1/4 in. NPT, 1/4 in. tube) (to vacuum pump)
- Female connector to NPT(1/4 in.), B-400-7-4RT

**Additional supplies:**

- Cork borers for cutting connector holes: Cole-Parmer, U-02698-90
- Snoop, for leak testing: Swagelok MS-SNOOP-8OZ
- Bound laboratory notebook for record keeping

## Protocols for Anoxia in Large-Scale Systems

Anoxia treatment on a large scale requires first flushing the air from the enclosure with nitrogen and then continuing a slow, steady flow of the gas to remove any oxygen that might leak in during treatment. Large-scale enclosures are 1 $m^3$ and larger.

Because of the large quantities of nitrogen involved, there must be ample room provided for the use of liquid nitrogen in very heavy dewars (Fig. 4.6, p. 65) or for 6 to 12 nitrogen gas cylinders on a steel pallet that requires a forklift for transportation (Fig. 4.3, p. 63). Such an arrangement is often called a cradle or a six-pack. Another nitrogen-supply alternative is a bulky nitrogen generator system (Fig. 4.8, p. 66), which also must be delivered by forklift. A specialized utility power, such as 220 V or 240 V, 3-phase, may be required for operating a nitrogen generator. An additional ventilated space may be needed for an heat exchanger when using liquid nitrogen.

Safety measures must also be planned to avoid nitrogen leaks during treatment that would endanger the health of workers. Precautions include installation of a low-oxygen alarm and assurance of standard, rapid air exchange in the treatment room or provision of a duct to exhaust nitrogen to outside the building.

The following protocols describe three variations of large-scale anoxia systems commonly used to treat infested museum objects: a large user-made tent (Protocol B); a large commercial enclosure known as a Portable Bubble (Protocol C); and a large, commercial rigid structure (Protocol D). Materials and suppliers are listed after each protocol, and a troubleshooting section for large-scale anoxia systems is provided at the end of this section.

## Protocol B: Anoxia in Large, User-made Tents

There is no reason why a large treatment tent cannot be made successfully in-house and function as well as a commercial unit. However, the construction and manipulation of a large, user-made system requires considerable forethought and preparation, especially regarding the physical location for the anoxic operation as described above.

The treatment system described below consists of a plastic-film tent over an internal support frame on a specially prepared base.

Work space is a major consideration when making the system. Fabrication of a 10 $m^3$ tent as shown in Figure 4.13 (p. 71), for example, requires a 4 m × 8 m clear floor space for optimal working conditions. Construction of the internal support frame needs about a 4 m × 4 m working area. Thus both the tent and its internal support frame cannot be built at the same time unless there is a moderately spacious work space available. An additional consideration is the treatment room's ceiling height, because the tent is raised off its base to allow for loading the objects to be treated. For a 10 $m^3$ tent, the ceiling should be about 6 m high.

Reviewing the space requirements above and following the step-by-step instructions below will assure successful fabrication and operation of a large anoxic enclosure. The instructions that follow are divided into two parts: Part 1, Enclosure Fabrication and Testing; Part 2, Anoxic Treatment.

### Part 1: Enclosure fabrication and testing

**Step 1 (prepare base):** The concrete floor on which the tent and its support frame will rest can be used as the base if it is level and smooth and has no visible cracks. It first must be sealed with several layers of epoxy paint to prevent any air from coming up through the floor and into the enclosure. The painted base should extend about a meter beyond the perimeter of the enclosure. The paint should be allowed to dry and outgas for roughly two weeks before treatment begins (follow the manufacture's recommendations for drying time).

**Figure 5.8, below**
Assembly of the lifting device for a 1 $m^3$ tent

**Figure 5.9, right**
The lifting device of the 1 $m^3$ tent is attached to a pulley system.

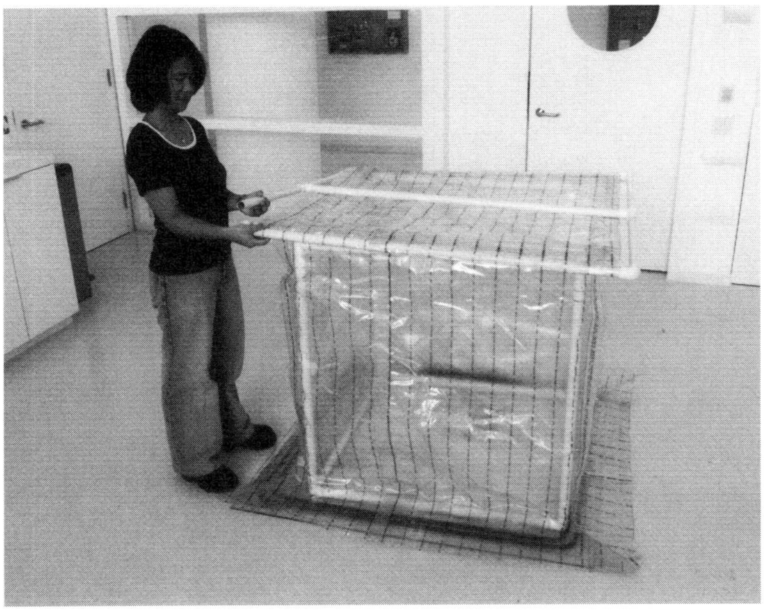

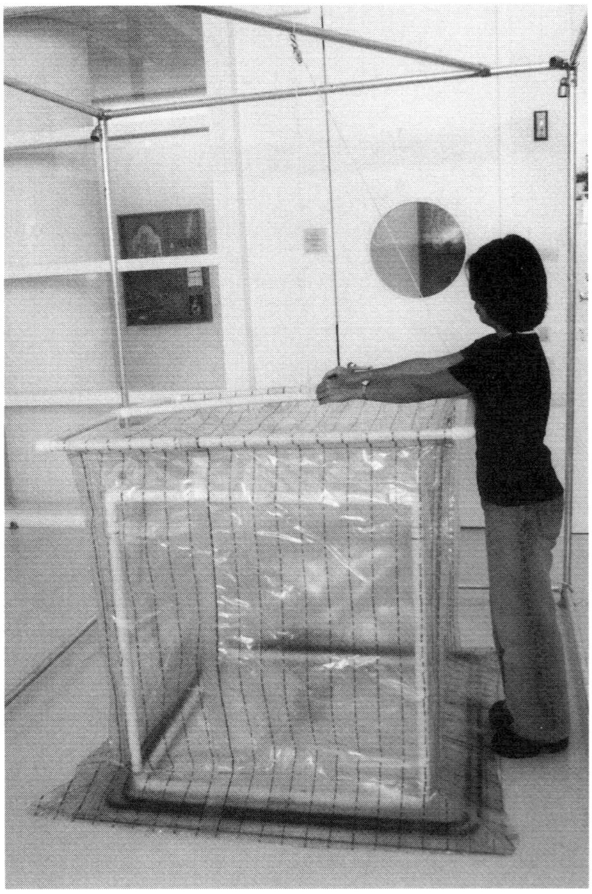

Alternatively, an acrylic sheet at least 6 mm thick can be used as the base. It should extend 20 cm to 25 cm beyond the perimeter of the enclosure. For example, a 10 $m^3$ enclosure with a 2 m square footprint can be accommodated on a 2.44 m square base. This can be made by joining two standard-size acrylic sheets (each 1.22 m × 2.44 m) with metal adhesive tape.

Sheets of 2 mm thick stainless steel or 3 mm thick aluminum can also be used for the base.

**Step 2 (fabricate lifting device):** A lifting device is used to raise the tent off its base when objects are added or removed. The lifting device is illustrated for a small (1 $m^3$) enclosure in Figures 5.8 and 5.9. Made of inexpensive white Schedule 40 PVC irrigation pipe, the device is adequate for this small enclosure, although it is somewhat flexible. It can be stiffened, as shown in the figures, by using tee pipe fittings and inserting a cross member. Cut the four pieces of pipe with a PVC pipe cutter. The finished length, including elbows and tee fittings, on each side of the lifting device should be about 5 cm to 10 cm longer than the top of the internal support structure (see Step 3).

For the larger 10 $m^3$ tent in Figure 4.13 (p. 71), a more rigid lifting device was made from 2.5 cm aluminum tubing and slip-on fittings. The aluminum tubing must be sawed or cut with a tube cutter, and any burrs should be filed so that the slip-on fittings will fit easily.

Leave all pieces of the lifting device unassembled until Step 6.

Prepare for the lifting device by securely mounting a sturdy pulley on a ceiling beam or girder capable of hoisting about 20 kg (for a 10 $m^3$ tent), centering the pulley over the enclosure base (Fig. 5.10). Run a rope through the pulley; one end will be attached to the lifting device that will be joined to the top of the tent. Wrap the other end of the rope around a cleat mounted on the nearest wall.

**Step 3 (fabricate internal support frame):** The framework supporting the plastic-film tent can be made using 2 cm PVC pipe for a 1 $m^3$ unit (Fig. 5.11) or using 2.5 cm aluminum pipe for a 10 $m^3$ unit. The more expensive aluminum tubing and fittings are superior, but they are not necessary for the smaller unit. When assembled, the PVC frame is already adequately rigid.

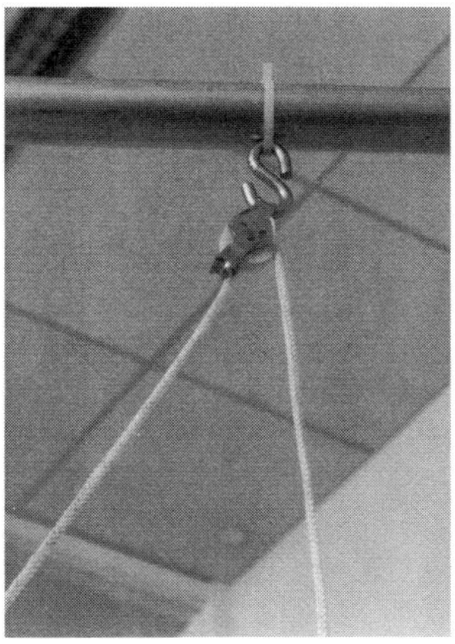

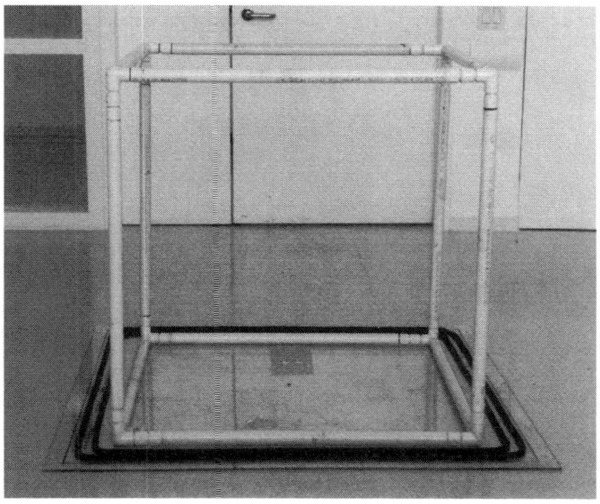

**Figure 5.10, right**
A pulley of the lifting device installed above a 1 $m^3$ tent

**Figure 5.11, far right**
Supportive frame for a 1 $m^3$ tent, made from PVC tubing

The finished support frame should be slightly smaller than the tent to make it easy to lift the tent off the frame. The finished frame should be approximately 5 cm to 10 cm shorter in width and length, and no less than 5 cm lower in height, than the tent. Cut the PVC pipe and temporarily fit the pieces together with elbows to double-check its size, and then complete fabrication of the frame. Center the frame on the base, being careful not to scratch the base.

At this time, shelving can be carefully placed within the frame if a number of small objects are to be treated. Placing objects on shelves allows for better circulation of the low-oxygen atmosphere. If metal shelving will be placed on a painted base, cover the base first with a heavy tarpaulin or corrugated cardboard sheet to prevent gouging that might lead to leaks. (Acrylic and metal bases are rugged enough not to need this protection.) The tarpaulin or cardboard remains in the unit throughout treatment.

**Step 4 (make the tent pattern):** The tent is fabricated out of oxygen-barrier film. A general pattern for any size tent is shown in Figure 5.12. A 10 m$^3$ unit would measure 2 m in length and width and 2.5 m in height. The back, top, and front panels are made from one continuous sheet of film. The shaded areas show where the film will be sealed according to the number sequence in Figures 5.13 and 5.14, although the sequence of seam sealing is the same.)

The pattern allows a generous 8 cm wide area for the seams so that optional double seams are easy to make within the 8 cm, and even a third corrective seam is possible. Although the front, top, and back panels are made from a continuous sheet of film, a seam is

**Figure 5.12, below**
Pattern for a tent, with areas to be sealed indicated by shading.

**Figure 5.13, right**
Front, top, and back panel of the tent (not to scale) with sample dimensions for a unit about 6.8 m$^3$. The heat sealing should be performed according to the sequence of the numbers.

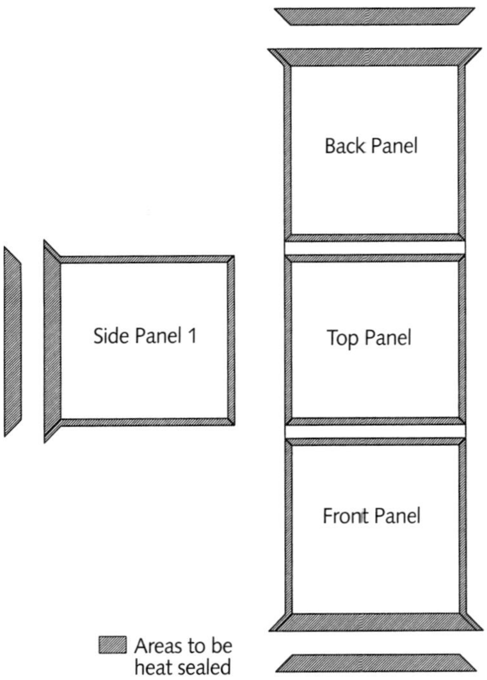


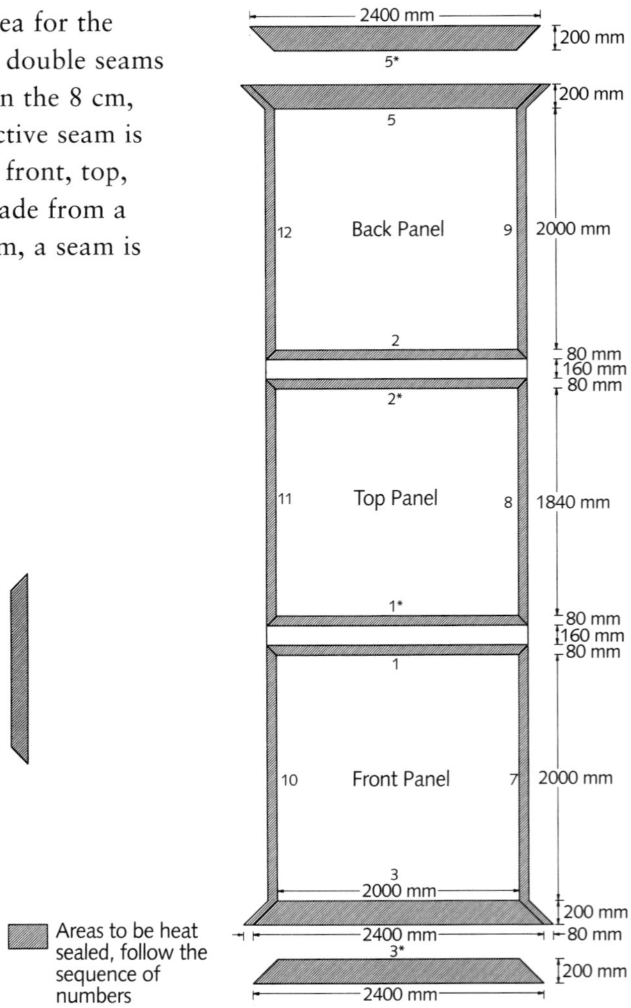

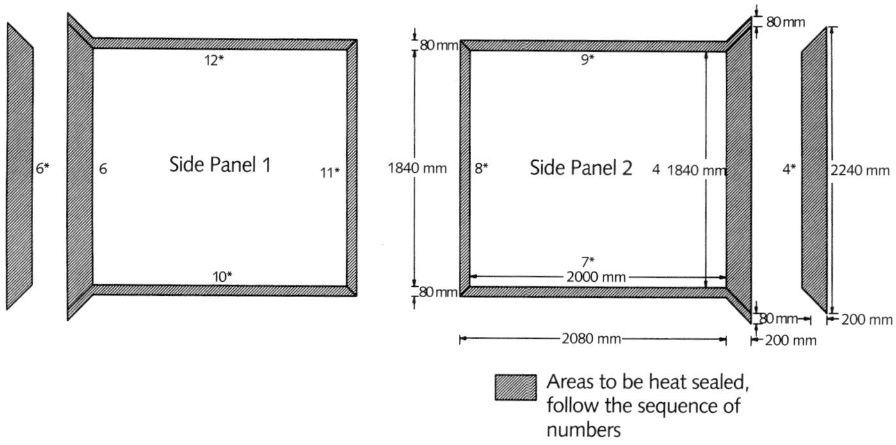


**Figure 5.14**
Side panels of the tent in Figure 5.13

still made where the top panel meets the front and back panels as indicated on the figures. This will create an open-ended tube along two sides of the tent top. The lifting device will be inserted through these loops. The shaded, trapezoidal pieces of film in Figures 5.12 through 5.14 will be sealed to the corresponding parts of the bottom edges of the tent to strengthen them and create a flange. The flange will be used to seal the tent to the base. The sides of the tent (see Fig. 5.14) will be the last to be sealed.

The pattern shown in Figure 5.12 can be proportioned to make any size enclosure. For a 10 $m^3$ tent, each seam should be at least 2 cm wide to allow for an optional double seam without crowding or being too near an edge.

**Step 5 (cut and assemble tent panels):** Use a rigid 120 cm (4 ft) metal carpenter's ruler and a permanent marker to accurately lay out the pattern on the selected oxygen-barrier film. Cut the film with scissors or with a sharp-bladed cutter.

On one of the panels, about 10 cm from the bottom near a corner, mark where the nitrogen inlet connector will be installed. Then mark the location of the nitrogen outlet connector; this should be diagonally opposite the inlet, 10 cm from the top corner. Both the inlet and the outlet will be fitted with Swagelok O-ring connectors.

Approximately 15 cm from the nitrogen inlet (either side) and 10 cm from the enclosure bottom, draw a small circle to mark the installation point for a hermetically sealed electrical connector that will allow a hermetically sealed set of wires to pass between RH and oxygen sensors inside the enclosure and their corresponding read-out devices on the outside (see "Electrical connectors" in Appendix I, p. 134).

Using patches, reinforce the film at points where heavy connectors will be located on the tent as follows: heat seal an 8 cm square of film centered over the mark for the nitrogen exit connector and a 25 cm × 8 cm rectangle covering the marks for both the nitrogen inlet connector and the electrical connector. Make a double seal along the midline of this last reinforcement, between the two connectors. See the discussion "Leak-proof Connectors in Anoxic Enclosure Walls," page 133, in Appendix I for additional information.

Use a sharpened cork borer just barely equal to the diameter of the threaded end of the Swagelok O-ring connector to make holes in the film where the nitrogen inlet and outlet connections are marked. First place a small piece of corrugated cardboard behind the connector mark on one side of the film for backing, then from the other side of the film slowly bore a precise hole for the connector. Do most of the cutting by rotating the borer rather than exerting great pressure and possibly forming a misshapened hole. The hole for the electrical connector will be larger than the hole for the Swagelok connector and does not have to be as precisely made. The hole can be carefully cut with a sharp Exacto knife, the preferred tool. If another kind of art knife is used, be careful not to create a misshapen hole that can lead to a leak. The size of the hole for the electrical connector should be smaller than the inner diameter of the O-ring on the connector parts shown in Appendix I, Figure I.3, page 128.

Once all the panels have been cut and the connector holes made and reinforced, heat seal all pieces of the tent together, following the number sequence indicated in Figures 5.13 and 5.14. Wrinkles in the long seams can be avoided by working on one or two long tables, such as a commercial 2 m folding utility table. This way a single seam can be made in a continuous strip, then more film can be carefully folded over on the table for the next seam, and so on. A hand-operated bar sealer is preferable to a tacking iron for long seals; it is not only faster but also more reliable because both sides of the film are heated together at once. However, a tacking iron may be needed for close work on some small corner seams and to repair any imperfect seams detected later during leak testing.

Next cut four 15 cm wide acrylic plates (at least 10 mm thick), long enough to cover the flange along the bottom of the tent (Figs. 4.14 and 4.15, p. 71). The plates should have perfectly squared ends so that they can mate closely.

**Step 6 (assemble treatment unit):**

(a) Place tent over internal support frame. With two or more persons assisting to avoid damaging the plastic film, carefully place the enclosure over its internal support frame made in Step 3. See Figure 5.15, which shows only one person because this is a small (1 $m^3$) enclosure.

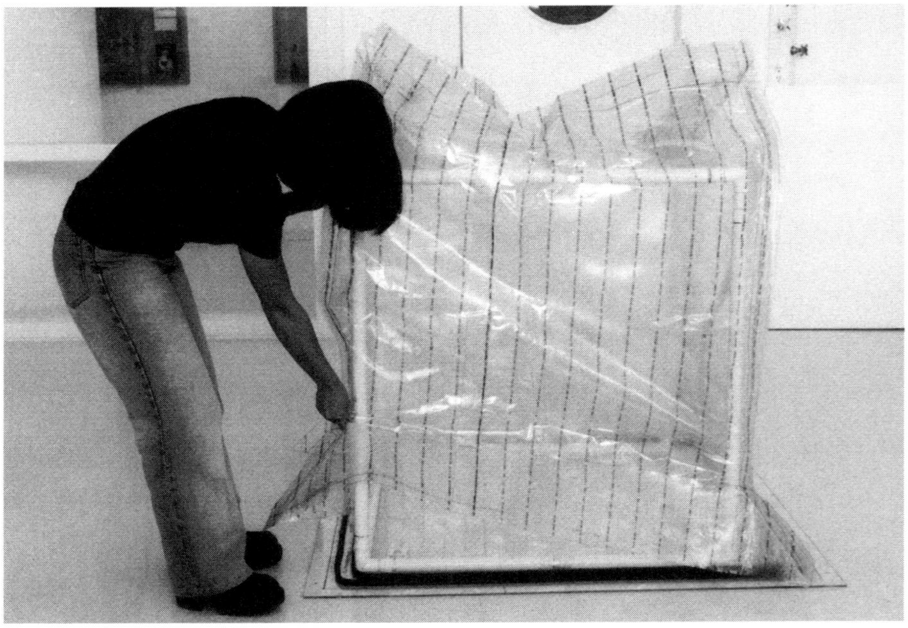

**Figure 5.15**

The 1 $m^3$ tent top is carefully placed cver the supportive framework.

(b) Attach lifting device. Slide two PVC pipes (or aluminum tubes) of the lifting device through the plastic-film tubes created along two edges on top of the tent. Then finish assembling the lifting device. The final assembly of a treatment enclosure with lifting device in place is shown in Figure 5.9 and in the schematic in Figure 4.14. (Note that the lifting device in the schematic is rotated 90° from the device shown in Fig. 5.9 and that it does not show a strengthening crossbar.) Attach the four corners of the lifting device by ropes to the pulley system as shown in Figures 4.14 and 5.9 (pp. 71, 98).

(c) Mark sealing point on base. The flange along the bottom edge of the plastic tent will be sealed to the base. The flange will rest over two rubber grommets (each a half-tube of rubber with adhesive on the flat side) and will be covered with the acrylic plates made in Step 5. These, in turn, will be weighted down with sandbags. In this step, draw a line around the inside perimeter of the base with a permanent marker, centering the line where the midline of the flange will be. This line will guide the later installation of the rubber grommets.

(d) Attach nitrogen inlet/outlet connectors. The O-ring connectors for the nitrogen inlet and outlets are simply inserted in the holes already prepared in the enclosure. Lightly greasing the connector's O-ring, washer, and nut will help prevent wrinkles from forming in the film around the hole as the nut is tightened. Properly tightening the nut is crucial because connectors can be the source of a major leak. The final turn with a wrench should be firm, but *not* extremely so, or the very thin O-ring will be overcompressed and possibly ineffective in sealing (Fig. 5.16). See the discussion "Leak-proof Connectors in Anoxic Enclosure Walls" in Appendix I for additional information on attaching connectors and supporting them so that they will not strain the plastic film and cause leaks.

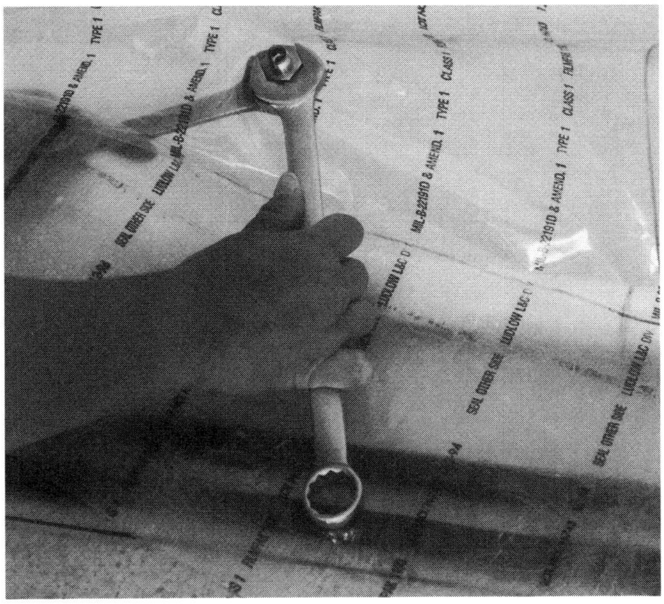

**Figure 5.16**
An O-ring connector is installed to be used as an entry or an exhaust port for nitrogen.

(e) Attach on/off valves using port connectors. In this step Whitey on/off ball valves are attached to the nitrogen inlet and outlet O-ring connectors via a port connector. *Note:* One end of the port connector has a completed vacuum fitting. A Swagelok nut needs to be added to the other end and tightened before attaching the port connector to the O-ring connector. This will minimize torque on the plastic film. The valve can then be attached, taking care to support both the valve and the O-ring connector during final tightening, again to avoid torque and twisting the film around the attachment hole. Support the finished valve unit to prevent stress on the film around the connector. During a treatment, be especially carefully when opening or closing valves attached to connectors on the tent. Hold the body of the valve firmly with one hand so that no torque is exerted on the connector when the other hand turns the valve handle.

(f) Install electrical connector. Use the same precautions described above when installing the electrical connector with hermetically sealed sensor wires (see Appendix I, Fig. I.8, p. 134). When done, temporarily attach the sensor wires to the internal support frame, either by looping the wires over the frame or attaching them with masking tape. (The wires will be attached later to sensors inside the treatment enclosure and to their corresponding monitors outside the enclosure.) In either case, the wires should be attached to the frame at a point 1 m to 1.5 m from the ground with a slack of 0.5 m to 1 m. This will make it possible to raise the tent off the support frame without stressing the electrical connector on the plastic film. Similarly, allow enough slack for the external wires that will be attached to monitors or dataloggers so there will be no strain on the tent or wires when it is raised or lowered.

(g) Make final preparations. Raise the tent to allow comfortable working space beneath it. Place two self-adhesive rubber grommets parallel to the guide line marked on the base in Step 6c. Each grommet should be about 1.5 cm from the guide line. Curve the grommets around corners (Fig. 5.17). Seal the ends of the grommets as follows: lay the end of one grommet over the other for a few centimeters, then, using a razor blade, cut diagonally through both layers; carefully remove the waste portions of each grommet and adhere the two diagonal ends together with Super Glue.

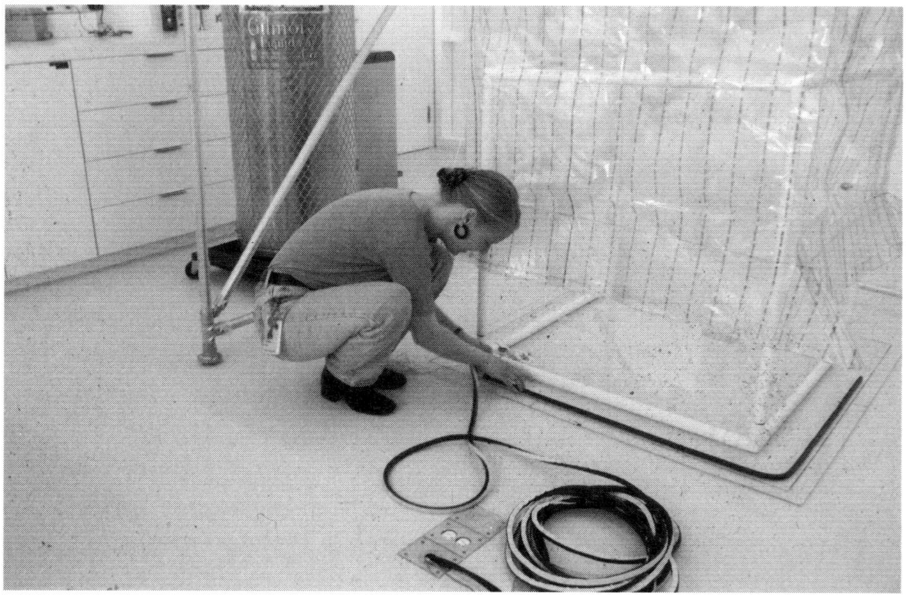

**Figure 5.17**

Self-adhesive rubber strips are placed on the floor, beneath the footprint of the flange perimeter.

**Step 7 (do pretreatment setup):** Place the objects to be treated within the perimeter of the support frame. If there are many small objects, put them on shelves rather than stack them.

Check to see that the rubber grommets are not contaminated with dust or grit, and, if necessary, clean them with a vacuum cleaner or with soap and water, drying carefully. Then apply a small amount of laboratory grease in a uniform film along the length of the grommets.

Lower the tent carefully so that its flange is in wrinkle-free contact with the greased grommets on all four sides. With the flange carefully smoothed in place, add the acrylic plates made in Step 5 to the top of the flange (Figs. 4.14, 4.15, p. 71). Fill slender bags with sand to make sand snakes and place them on the acrylic plates. This will exert uniform pressure on the flange, sealing it perfectly to the grommets.

**Step 8 (calibrate sensors):** Connect the color-coded electrical wires from the oxygen sensor in the enclosure to its monitor(s) outside. Calibrate the oxygen sensor and monitor against air, following instructions supplied by the unit's manufacturer. This will ensure accurate oxygen readings inside the enclosure during the anoxic treatment. Connect the proper wires from the RH sensor in the enclosure to the RH monitor or datalogger outside. Calibrate the RH sensor beforehand, either by comparing its reading with that from a calibrated sensor's reading or by placing the sensor in a chamber that has been RH-conditioned with a saturated salt solution.

**Step 9 (complete system setup and equalize nitrogen flow rate):** The schematic drawing of a large-scale treatment system in Figure 4.1 (p. 59) should be studied carefully before continuing with this step.

First, attach the nitrogen humidification system (the RH-conditioning module on the figure) to the tent enclosure with nylon tubing, but do not turn on the nitrogen supply. Next, it is important to know the rate of nitrogen flow into the enclosure so that it can be exactly matched by the draw of the vacuum pump attached to the nitrogen outlet to avoid building up pressure in the enclosure. This can be done by observing the rate registered by the flowmeters attached at the enclosure's nitrogen inlet and outlet. To facilitate this simultaneous reading, the two flowmeters should be placed near each other. To do this, make the tube leading from the outlet valve long enough so the outlet flowmeter (and its attached pump assembly) can be placed next to the inlet flowmeter on the other side of the enclosure. This is *not* shown in Figure 4.1.

The nitrogen outlet on the enclosure is connected by tubing to an *outlet valve* and then to a vacuum pump assembly with its own *pump control* valve. This assembly is then attached to the *outlet flowmeter*. The vacuum pump assembly is set up in a control loop that governs how much nitrogen is drawn from the enclosure per minute. During treatment, when the outlet valve is wide open and the pump control valve is closed, the pump will draw 31 L of nitrogen per minute from the enclosure as registered on the flowmeter. However, when the control valve for the pump is gradually opened, the amount of nitrogen recirculating in the pump assembly increases, and less nitrogen is drawn from the enclosure. The outlet flowmeter then shows that less that 31 L of nitrogen per minute is being drawn from the enclosure, and that rate becomes less and less as the pump's control valve is opened more and more. (At the extreme end, with the control valve wide open, no nitrogen will be sucked from the enclosure at all because it will simply recirculate in the pump assembly loop. If the outlet valve on the enclosure is closed, the pump will do exactly the same thing, running at maximum in an endless recycle.)

Therefore, closing the pump control valve to various degrees precisely controls how much nitrogen the pump draws from the enclosure. The inflow and outflow of nitrogen needs to be precisely controlled throughout the anoxic treatment.

In completing the final system setup, make sure that all tubing to the inlet and outlet valves is long enough and well supported so that it does not stress the plastic-film tent at the connector points.

**Step 10 (leak test the enclosure):** Before proceeding, read the instructions for leak testing of anoxic enclosures in Appendix I.

Prepare for leak testing as follows: (a) close the nitrogen outlet valve on the enclosure; (b) detach the inlet flowmeter/relief valve tubing from the nitrogen inlet valve; (c) open the nitrogen inlet valve. (Remember: when attaching or detaching connections or turning valves, do so in a way that does not exert any torque on the connector and its attachment point to the plastic-film enclosure.)

The leak test is conducted in the following sequence:

(a) Inject tracer gas into the inlet valve (less than a 5 second burst of tracer gas for every 1 $m^3$ of volume) and then close the valve (Fig. 5.18). Reconnect the inlet flowmeter/relief valve tubing to the inlet valve.

(b) With the nitrogen outlet valve still closed, open the enclosure's nitrogen inlet valve and close the relief valve located near the inlet flowmeter. Turn on the nitrogen source and allow nitrogen to flow through the humidification unit and into the treatment enclosure at an arbitrary rate, then increase the rate to 30 L/min for a enclosure that is 0.5 $m^3$ or larger. By keeping the nitrogen outlet valve closed at this point, the in-flowing nitrogen will pressurize the tent slightly. This will increase the tent's tendency to leak and thus make it easier to find defective seals or other causes of leakage.

(c) Before leak testing, adjust the valves on the humidification unit's two flowmeters (see Fig. 4.10, p. 68) to achieve the desired RH level for the objects to be treated, using data previously logged during construction and testing of the humidification system (see Appendix I). After a minute or two, record the reading from the RH sensor inside the anoxia tent.

(d) When the walls of the anoxia tent begin to bulge slightly from the nitrogen buildup, do the following: (i) open the relief valve between the inlet flowmeter and inlet valve; (ii) close the inlet valve; (iii) quickly close the nitrogen entry valve to the humidification system. Then follow the instructions for leak testing of anoxic enclosures in

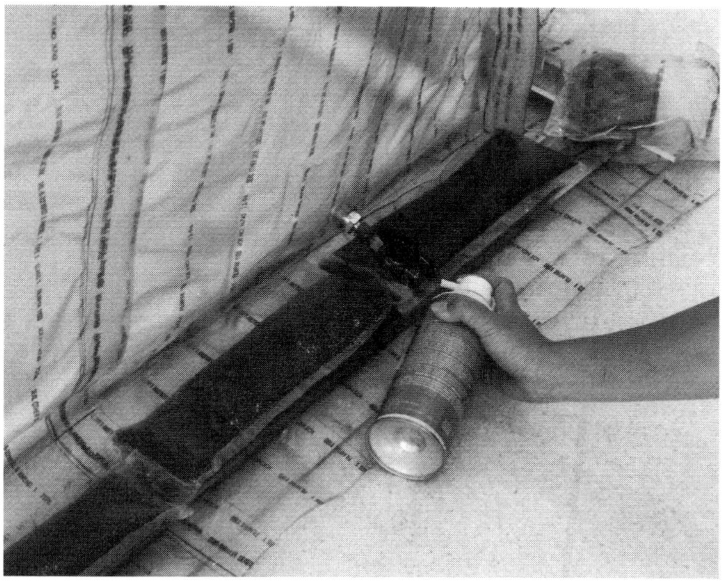

Figure 5.18

A small amount of tracer gas is injected into the enclosure through the nitrogen inlet port.

Appendix I. Check for leaks with the halogen leak detector, especially along the seal around the flange at the bottom of the enclosure, around connectors, and along all heat-sealed seams. Correct any leaks by resealing the seam. If the leak is caused by a wrinkle in the film, make a second seal around the leak point (resealing wrinkles usually fails). Repeat the leak check 15 minutes after fixing leaks until no more leaks are detected.

**Part 2: Anoxic treatment**

Most insect species common to North America and other countries are killed at an oxygen level below 0.3% (optimal range is 0.2% to 0.1%); an RH of 50%; and a temperature of 25°C. The temperature inside the treatment enclosure is controlled by the temperature of the room in which the enclosure is housed. The standard treatment lasts fourteen days, which includes two days for purging the system to reach the desired oxygen level. (A twenty-two-day treatment is advisable where *Hylotrupes bajulus* (old house borer) and other anoxia-resistant insects are endemic.)

**Step 1 (purge air from enclosure):**

(a) Assemble the entire treatment system as per Figure 4.1 (p. 59), then adjust the various valves as follows:

- *nitrogen source:* Check that the valve from the nitrogen source to the humidification unit is closed. If using a nitrogen cylinder, turn the diaphragm handle on the cylinder clockwise until the gauge reads about 40–50 psig (3–3.5 barg). If using liquid nitrogen, adjust the delivery pressure similarly. If using a nitrogen generator, turn it on, and when the nitrogen production is stable at 99% purity or better, open the generator's valve to the humidification unit.
- *humidification unit* (see Fig. 4.10, p. 68): Make sure the valve on the wet flowmeter is open; the pressure relief valve on the water bottle is closed; and the valve for adjusting RH on the wet line is opened at least one turn.
- *treatment enclosure:* Open the relief valve between the inlet flowmeter and the inlet valve, and then open the inlet and outlet valves on the enclosure. Turn on the vacuum pump with its control valve set so the outlet flowmeter registers only 5 L/min to 10 L/min.

(b) Partly open the valve from the nitrogen source to the humidification system. Raise the nitrogen in-flow rate gradually to 30 L/min as measured by the enclosure's inlet flowmeter. Adjust the ratio of wet to dry flow rates for the humidification system to the appropriate final flow rate, using the settings for the flowmeter valves found during calibration of the humidification system in Appendix I. While the flow rate is increasing, check again that the enclosure's nitrogen inlet and outlet valves are wide open. Then close the relief valve. Moderately quickly, adjust the vacuum pump control valve so that the rate of nitrogen being pumped out of the enclosure, as shown on its outlet flowmeter, is exactly the same as the rate flowing into it, as shown by the inlet flowmeter. An exact balance is impossible in the first few minutes, but once obtained, the balance between the nitrogen inlet and outlet flow rates must be maintained over the course of the anoxic treatment. This will prevent the walls of the enclosure from bulging and stressing heat-sealed seams or contracting and disrupting the seal between the flange at the bottom of the tent and the grommets on the base.

***Caution:*** During the purging process, nitrogen is being exhausted out of the treatment system. To prevent a dangerous drop in the oxygen level of the room housing the treatment tent, either turn on the room's forced ventilation system or vent the exhaust to the outside.

Continue monitoring the enclosure's two flowmeters, adjusting the pump control valve as necessary to maintain equal rates. Ensure that the nitrogen flow rate into and out of the enclosure stays constant at 30 L/min and that the RH remains at 50% (when adjusting the valves on the humidification system, it takes minutes to hours to see the effect on RH in the enclosure).

(c) Check the readings from the RH and oxygen sensors inside the enclosure about every fifteen minutes for the first hour or two. Once the relative humidity is stabilized, the time between observations can be extended to hours, then to twice a day, and finally to daily. If the RH and oxygen values are not being continuously monitored using a datalogger, manually record the conditions inside the enclosure.

(d) The goal of this purging is to bring the enclosure's oxygen level to between 0.2% and 0.1%, the optimal range for insect anoxia. The amount of nitrogen required to purge an enclosure to this oxygen range is approximately 8 times the volume of the enclosure. The time it will take to purge the enclosure is approximately two days; it can be estimated from the flow rate for the volume of the enclosure. When a 0.2% to 0.1% oxygen level has been reached, the nitrogen supply to a well-sealed enclosure is shut off, and the treatment will continue by itself for the remaining twelve days before the oxygen level rises to an ineffective level (above 0.3%).

Shut down the nitrogen flow in the following sequence: (i) quickly open the relief valve near the enclosure's inlet valve; (ii) shut the enclosure's outlet valve; (iii) close the inlet valve; (iv) close the valve between the nitrogen source and the humidification unit; (v) close the valve on the nitrogen source itself; (vi) open the relief valve on the humidification system's water bottle; and (vii) turn off the vacuum pump. All valves should be closed at the end of this shut-down operation.

**Step 2 (repurge if necessary):** If during treatment the oxygen level rises about the optimum treatment level (0.2% to 0.1%), repeat the purge described in Step 1. However, if it is too difficult to maintain this low oxygen level in the enclosure, then a slow, steady flow of nitrogen is needed throughout the treatment process to offset the oxygen leaking into the system.

**Step 3 (terminate anoxic treatment):** After fourteen days the treatment is complete, and the anoxia tent is raised. In doing so, however, care must be taken not to inadvertently release a large amount of nitrogen into the workroom, which could drop the oxygen level to below 19.5% and endanger workers. A few seconds' exposure to 19.5% oxygen is not a cause for concern. Mountain climbers constantly work at this level. Nitrogen itself is not toxic; it is oxygen deprivation at 19.5% or below for a minute or more that is not advisable for ordinary people.

An oxygen monitor with an alarm set to go off when the oxygen level drops to 19.5% must be installed in the room where the treatment is taking place. It is highly advisable that individuals also carry such a monitor on their persons when opening an enclosure after treatment.

The amount of nitrogen in a 10 $m^3$ enclosure, if uniformly distributed throughout a totally unventilated room 8 m × 8 m with a 5 m ceiling, would cause a trifling drop in oxygen concentration—to 20.3%, a safe level. However, a dangerous situation can arise if the enclosure is quickly raised. The nitrogen does not mix instantly with all the air in the room. Consequently, the oxygen level near the enclosure could easily drop far below 19.5% for an appreciable time, even in a larger well-ventilated room, endangering anyone standing nearby. See "Safe Use of Large Volumes of Nitrogen" in Chapter 4, pages 61–62, for a detailed discussion of nitrogen hazards.

Two simple procedures to avoid such a hazard when terminating an anoxic treatment are described below, along with an alternative approach.

(a) The most convenient procedure is first to remove the sandbags and acrylic plates weighing down and sealing the flange on the treatment tent to the enclosure base. Then raise the tent by about 50 cm to 1 meter, fasten the pull rope on the wall cleat, and leave the room for thirty minutes (if the room has mechanical ventilation) or for an hour with the windows and doors open (if the room does not have mechanical ventilation). After enough time has passed, the tent can be raised to full height, but *do not* immediately enter the treatment space itself and begin to remove objects. Leave the room again for fifteen minutes to allow for total mixing of the nitrogen with the room air before returning and removing the treated objects. Of course, the oxygen-monitor alarm is the best indicator of an unsafe oxygen level.

(b) The second procedure to dissipate the nitrogen is to use the vacuum pump on the treatment system to remove the gas from the treatment enclosure while drawing air in the entry valve. Do this in the following sequence: (a) open the enclosure's inlet valve and the adjacent relief valve; (b) make sure the relief valve on the humidification unit is also open; (c) open the enclosure's outlet valve, if it is not already open; and (d) close the control valve on the vacuum pump. Then turn on the pump and let it run for about fifteen hours, exhausting the nitrogen to outside the building. This arrangement forces the pump to exhaust a maximum flow of nitrogen from the enclosure. With about 30 L/min of air being drawn through a 10 $m^3$ enclosure for that length of time, this would amount to 2.7 volume changes, and the oxygen concentration inside the enclosure should be about 17%. At this point, the enclosure can be safely raised completely, but personnel should still leave the room with windows opened for half an hour before returning to work in the treatment area. The personal low-oxygen monitor should still be worn.

**Alternative:** A standard tank-type vacuum cleaner can be used to suck the nitrogen from the treatment enclosure and exhaust it to outside the building. Attach the vacuum cleaner's hose to the enclosure's outlet valve with duct tape, *but support the hose so its weight will not stress the enclosure at the connector site.* A slightly neater arrangement is to attach a Swagelok B-400-R-12 reducer, with a ¼ in. to ¾ in. adapter via port connectors, to the outlet valve. The ¾ in. tube can then be duct-taped to the vacuum cleaner hose end. In either arrangement, another vacuum cleaner hose is attached to the output of the cleaner and then to others, if necessary, until the hose extends to the outside the building.

The next step is to open the treatment enclosure's relief valve and inlet valve and turn on the vacuum cleaner. The process can be greatly speeded up by removing the sandbags and acrylic plates from the flange at the bottom of the treatment tent and raising the tent 25 cm to 50 cm off its base while the vacuum cleaner is running, again making sure that the hose is supported and places no stress on the outlet valve and connector. In an 8 m square room with a 5 m ceiling, it will be safe to turn off the vacuum cleaner after an hour. Untape the hose and raise the enclosure to its full height. Personnel must still wear an oxygen-monitor alarm for safety before removing objects from the treatment area.

## Materials and equipment

See Appendix IV for supplier contact information.

**For enclosure fabrication:**

- Oxygen-barrier film, preferably transparent and with an oxygen transmission rate below 51 cc/ $m^2$/day (tested in 100%

oxygen). See "Heat Sealing of Barrier Films to Form Anoxic Enclosures" in Chapter 2; Table 2.3 for guidance in choosing the appropriate film; and Appendix IV for suppliers.
- Tacking iron, Teflon coated, temperature regulated: Conservation Support, local photographic store, McMaster-Carr 1983T2
- Bar sealer, temperature regulated, or impulse bar sealer: Conservation by Design, IMPAK, Metode, McMaster-Carr 2064T16
- Carpenter's 4 ft rigid metal ruler; permanent marker; Scotch tape, masking tape, self-adhesive metal tape, wire; scissors, sharp knife, Exacto-type knife

**For internal support frame and lifting device:**
- Plastic PVC pipe and fittings (elbows, tees, corners) or aluminum tubing and fittings: local suppliers, McMaster-Carr
- PVC pipe cutter, if PVC pipe is used, or aluminum pipe cutter
- Pulley, wall cleat, plastic or conventional rope

**For enclosure base:**
- Epoxy paint to seal floor, or acrylic sheet (minimum 6 mm thick), to cover a 2.44 m square area: McMaster-Carr, self-adhesive rubber strips, 3/8 in. (10 mm), brown McMaster-Carr 93705K34
- Vacuum grease or laboratory grease or Vaseline: Cole-Parmer, U-79751-30 or U-17780-02
- Acrylic plates, 10 mm thick and about 15 cm wide, to cover flanges along the bottom of the tent; length according to enclosure dimensions
- Narrow bags filled with sand to weigh down the acrylic plates on the flange at the bottom of the tent

**For nitrogen supply:**
- Liquid nitrogen (with heater, internal or external); or
- Nitrogen cylinders (six mounted in a cylinder bank, six-pack, or on cradle) and attached to manifold that leads to enclosure; or nitrogen generator (membrane type): Air Products, Permea Division, PRISM® Models 1100 or 1300
- Nitrogen two-stage regulator, with gauges: Cole-Parmer, U-98202-23
- Nylon tubing, 1/4 in.: Cole-Parmer U-06489-22 (specified for use in assembling humidification system, but also needed for leading nitrogen to bag)
- Copper tubing: Swagelok CU-T4-S-050-50 (1/4 in., 50 ft) with fittings to connect regulator to humidification unit

**For humidification unit:**
- See list of supplies in Appendix I, "Construction and Use of a Nitrogen Humidification System."

**For leak testing:**
- See instrument specification for halogen leak detector in Appendix I, "Leak Testing of Anoxic Enclosures."

**For oxygen monitors (alarm set for 19.5% oxygen or below):**

- Oxygen monitor, mounted at head height near enclosure, with alarm: GC Industries
- Portable personal oxygen monitor with alarm: GC Industries

**For nitrogen removal from enclosure (alternative method):**

- Vacuum cleaner, tank type, local household type with Swagelok fitting to connect nitrogen exit port to vacuum cleaner hose, B-400-R-12 or B-600-R-12
- Duct tape

**For connections to enclosure from humidification system and from enclosure to vacuum pump, and to electrical sensors inside enclosure to monitors or dataloggers outside:**

- 2 Swagelok O-ring straight thread male connectors, brass. B-400-1-OR (with $^1/_2$ in. hole washers, $^7/_{16} \times 20$ in. nuts)
- 2 Whitey ball valves, brass, B-42S4 ($^1/_4$ in.)
- 1 Whitey needle valve, brass, B-1RS4 ($^1/_4$ in.)
- 5 port connectors, brass, B-4012-PC ($^1/_4$ in.)
- 2 union tees, brass, B-400-3 ($^1/_4$ in.)
- 4 female elbows (for flowmeters), B-400-8-2 ($^1/_8$ in. NPT, $^1/_4$ in. tube)
- 2 male connectors, B-400-1-4, ($^1/_4$ in. NPT, $^1/_4$ in. tube) for pump to tubing connection
- 2 flowmeters with valves (4–40 L/min), Cole-Parmer U-03216-38
- Nylon and copper tubing, $^1/_4$ in.
- Electrical connector with hermetic seal, fabricated from polypropylene cylinder by machinist (see Appendix I, "Leak-proof Connectors in Anoxic Enclosure Walls," and Figs. 5.10, I.4, I.5, pp. 99, 130, 133)
- Artist's sharp knife, such as an Exacto (for cutting hole for electrical connector)
- Cork borers: Cole-Parmer U-06298-90
- Vacuum pump, diaphragm, 31 L/min capacity: Cole-Parmer, U-07061-20

**For RH, temperature, and oxygen monitors:**

- RH monitor: Cole-Parmer U-37100-15, Conservation Support
- RH/temperature datalogger: Cole-Parmer U-80080-10, Conservation Support
- Electronic oxygen monitor:
  - (a) small: GC Instruments monitor; or
  - (b) sensor plus external monitor: Teledyne 320B/RC (with remote sensor) or Teledyne 3190 (trace analyzer with remote sensor); or
  - (c) head-space analyzer: hand-held head-space analyzer, PBI-Dansensor Checkpoint; movable head-space analyzer with printer: PBI-Dansensor Checkmate

- Datalogger to collect data from RH, temperature, and oxygen monitors: Conservation Support

**Miscellaneous supplies:**
- Cardboard or tarpaulin to protect painted base
- Shelving for objects

## Protocol C: Anoxia in Large Commercial Enclosures (Portable Bubbles)

Large commercial enclosures, or Bubbles, are described in Chapter 4, in the discussion "Tests of a Commercial Fumigation Bubble for Museum Anoxia." These cubical bubble enclosures are manufactured by Power Plastics in the United Kingdom. The single-use, Controlled Atmosphere Technology (CAT) Bubble is made from an aluminized polyester-PVDC laminate. The CAT Bubble comes with its own separate base but not an internal support frame. Though relatively inexpensive (from $110 for 1 $m^3$ to only $260 for 27 $m^3$), the CAT Bubble's lack of reusability decreases its attractiveness unless the conservator has only an occasional need for anoxia and cannot take the time to make an enclosure.

Power Plastics also makes durable and reusable PVC Portable Bubbles, similar to the 6 $m^3$ unit that was tested at the GCI and found to be satisfactory (see Chapter 4). However, the moderately high oxygen leak rate would mean a nitrogen cost of about $90 per treatment even with liquid nitrogen.

The instructions below, which are for a reusable Portable Bubble, are similar to those for the user-made tent described in Protocol B. Thus there are frequent references to that protocol. The same considerations apply regarding space to accommodate the nitrogen source and the installation of a low-oxygen alarm for safety. A Portable Bubble requires less setup space than a user-made tent, since it is already fabricated and is shipped folded into relatively compact units.

The manufacture's instructions for the Portable Bubble should be read before proceeding.

**Step 1 (prepare site for Portable Bubble):** The site where the Bubble will be located must be flat, clean, and free of any debris that might pierce the base and cause a leak. Cover the floor with a tarpaulin large enough to protect the Bubble base from abrasion, then unfold the base of the Bubble and center it on the tarpaulin. Cover the interior of the base with another tarpaulin or cardboard to prevent damage while the rest of the Bubble is being set up and the objects to be treated are added. The tarpaulin/cardboard will remain in the unit during treatment. Make sure that the tarpaulin/cardboard does not interfere with sealing the Bubble to the base.

**Step 2 (fabricate internal support frame):** Follow the instructions in Protocol B (Part 1, Step 3) for fabricating the Bubble's internal support frame. The only departure is that the frame should have a cross member on the top to hold the Bubble, which is folded up and stored on top of the frame when objects are removed after an anoxia treatment. Tie a rope across the top of the frame, perpendicular to the cross member

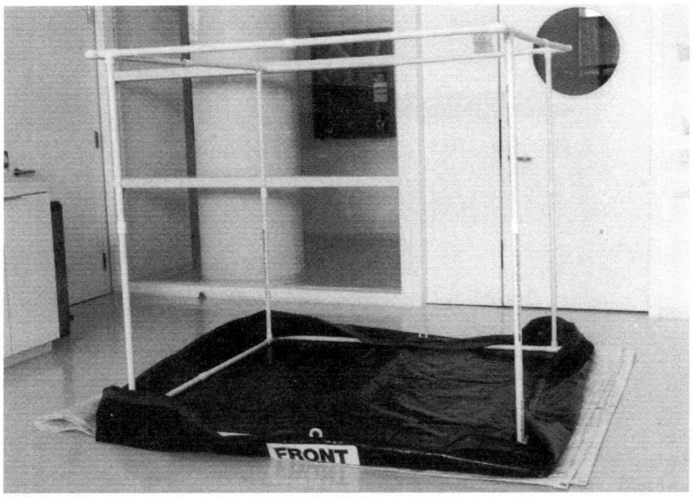


Figure 5.19

Supportive frame for the 6 m³ Portable Bubble, made from PVC tubing and fittings

and looped around it. This rope will be used to secure the folded Bubble so that it will not accidentally fall down after treatment is terminated. Place the support frame over the base. The support frame built for the 6 $m^3$ Bubble tested at GCI is shown in Figure 5.19, but without the cross member on top.

**Step 3 (install connectors on Portable Bubble):** The Portable Bubble comes with two built-in plastic ports with screw-on caps to which Swagelok O-ring connectors will be attached to create the nitrogen inlet and outlet. These ports are accessible on the folded Bubble, which should be placed next to the support frame when the following connections are made.

*Nitrogen connectors:* Remove the port caps and drill a hole in each with a 7/16 in. brad point drill that matches the threaded end of the O-ring connectors, then attach the connectors to the port caps with nuts (Fig. 5.20). Screw the caps tightly into the inlet and outlet ports, making sure that the gaskets in the caps are correctly positioned. Use Swagelok nuts and port connectors to attach two Whitey ball valves to the O-ring connectors.

*Electrical connector:* About 15 cm in any direction from the port cap that will be the nitrogen inlet, carefully cut a circular hole in the Bubble with an Exacto knife to accommodate the electrical connector that has been hermetically sealed with 2 m or longer color-coded sensor wires. The size of the hole is determined by the custom-made connector. Install the electrical connector with care to avoid wrinkling the plastic-film wall of the Bubble. For guidance, see the discussion on leak-proof connectors to anoxic enclosures in Chapter 4 and in Appendix I, as well as Figure I.5 (p. 133). Attach the color-coded wires to the RH and oxygen sensors and place the sensors inside the support frame, enclosing them in a small plastic bag to protect them later when the plastic Bubble is installed. Loosely loop the wires over the frame to allow enough slack so that they do not stress the connector site when the Bubble is installed. At this point, the external wires can be attached to the appropriate monitors/dataloggers.

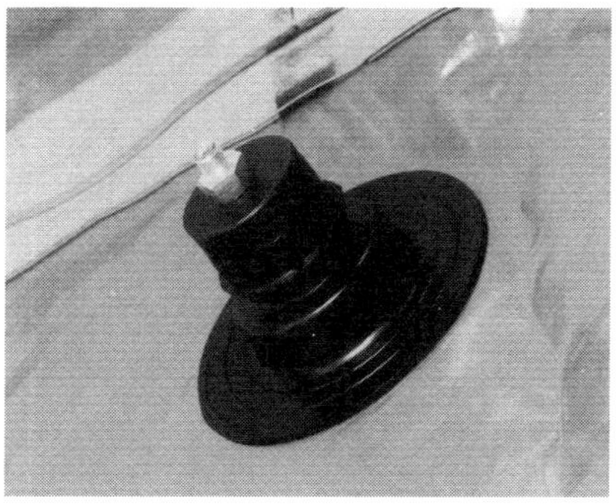

Figure 5.20

An O-ring connector is installed on a Portable Bubble port to be used as a nitrogen inlet or outlet.

**Step 4 (add infested objects):** Place the objects to be treated inside the support frame. Make sure they are not placed too close to the frame boundary to avoid having the heavy plastic Bubble accidentally touch them when it is unfolded over the frame and when it is folded up after treatment.

**Step 5 (install Portable Bubble):** With the aid of one or more people, depending on the size of the Bubble, move the folded-up Bubble over the support frame, rest it on top, then slowly unfold it over the frame (Fig. 5.21). Be careful not to touch the objects or the sensors.

The bottom edge of the Bubble and the perimeter of its base each contain one-half of a plastic zip-lock that will be pressed together to make an airtight join between the Bubble and base. Make sure the zip-lock is clean and totally free of dust and debris; vacuum it as an

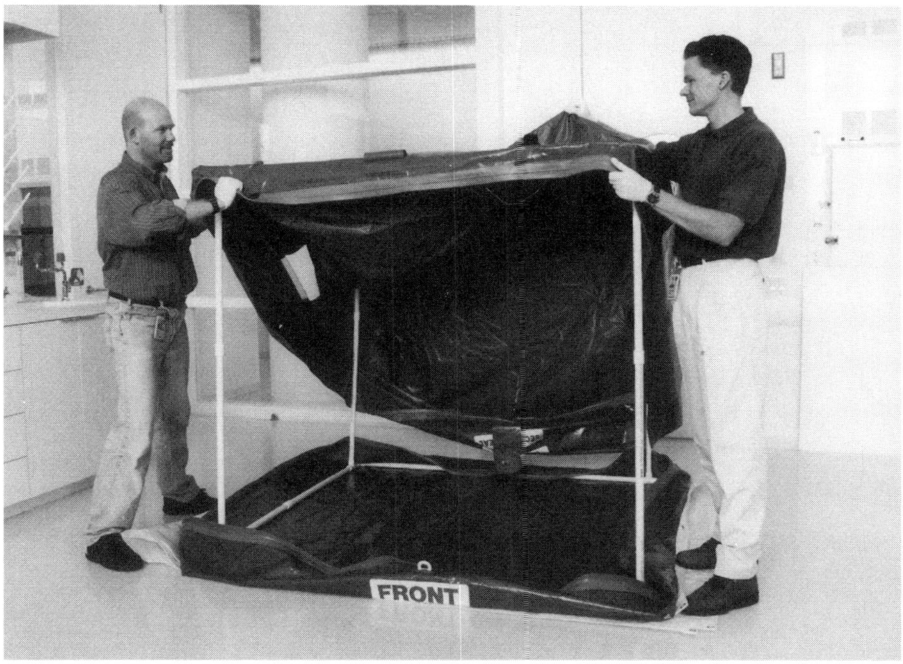


**Figure 5.21**
The Portable Bubble top is placed over its supportive frame.

extra precaution. Then apply a thin but sufficient coat of vacuum grease to one side of the zip-lock. Avoid getting grease on the objects, tools, shoes, Bubble wall, and so on. Colored markers on the outside of the Bubble indicate how the zip-lock halves line up. Using the colored markers for guidance, carefully close the zip-lock by applying sufficient pressure—or better yet, with the special tool supplied with the Bubble—to ensure a tight seal (Fig. 5.22).

**Step 6 (set up and calibrate final components):** At this point, the oxygen monitor is calibrated; the humidification system is attached and the nitrogen flow equalized; and the Bubble is leak tested. The instructions are the same as for the large user-made tent described in Protocol B, Part 1, Steps 8 to 10.

**Step 7 (purge air from Bubble and begin treatment):** The instructions for this step are identical to the ones for a large user-made tent. See Protocol B, Part 2, Steps 1 and 2.

**Step 8 (terminate treatment):** Read the introductory comments for Protocol B, Part 2, Step 3, then follow one of the two termination procedures below.

**Figure 5.22**
Closing the lightly greased zip-lock that holds the Portable Bubble top to its base

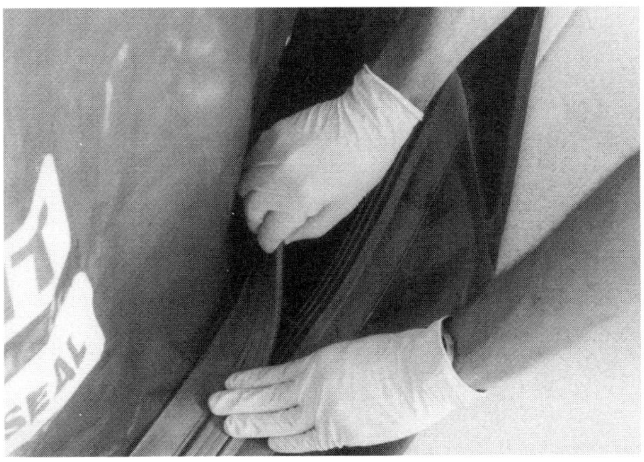

(a) The simplest procedure for removing the nitrogen safely from the Bubble is to let it diffuse into the empty workroom. First, open the windows and doors if the room is not mechanically ventilated. Then, with workers wearing a low-oxygen monitor with an alarm, disconnect the nitrogen inlet and outlet tubing from their valves, open the valves, and unzip the Bubble from its base. Use four small boxes, about 20 cm to 30 cm high, to prop up the walls of the Bubble, taking care not to touch the objects inside the support frame. Leave the room for one hour if it is mechanically ventilated; three hours if not.

At least two people, still wearing the low-oxygen monitor with alarm, should return to the room and carefully fold the sides of the Bubble up to the top of the support frame and tie the bundle down with the rope that was attached to the frame top in Step 2. If the alarm sounds as this lifting is being done, indicating a dangerously low oxygen environment in the room, the workers should leave the room immediately for at least an hour.

(b) The second procedure uses a vacuum pump as described in Protocol B, Part 2, Step 3. For the alternative using a vacuum cleaner: attach the hose to the nitrogen valve (either inlet or outlet) that is uppermost in the Bubble. This will cause air to flow into the Bubble from under its propped-up sides and to exhaust through the vacuum cleaner. More time is allowed for this "vacuuming"—two or three hours—than for airing the user-made tent in Protocol B. This is because the user-made tent can be completely raised off the floor, allowing more air circulation. More care must be taken when airing out a Bubble because workers must get close to it to prop up the sides, thus risking exposure to a low-oxygen condition. The user-made tent is raised at a distance using the rope and pulley, and the risk is not as great.

## Materials and equipment

See Appendix IV for supplier information.

**For the Bubble enclosure:**

- Portable Bubble, to 90 $m^3$: Power Plastics
- CAT Bubble, 1 $m^3$ to 27 $m^3$: Power Plastics
- Tarpaulin and cardboard to protect the base
- Vacuum grease or laboratory grease or Vaseline for zip-lock seal: Cole-Parmer, U-79751-30 or U-17780-02

**For internal support frame:**

- PVC pipe and fittings, or aluminum pipe and fittings: local suppliers, McMaster-Carr

**For nitrogen and electrical connections to the Bubble:**

- Brad point drill bits, 7/16 in. to bore holes in Bubble port caps
- Power drill, 3/8 in.
- 2 Swagelok O-ring straight thread male connectors, brass, B-400-1-OR (1/4 in., with 1/2 in. hole washers, 7/16 × 20 in. nuts)
- 2 Swagelok port connectors, brass, B-401-PC (1/4 in.)
- 2 Whitey ball valves, brass, B-42S4 (1/4 in.)
- Polypropylene block to machine an hermetically sealed electrical connector

**For Bubble sensors:**

- RH monitor: Cole-Parmer U-37100-15, Conservation Support
- RH, temperature datalogger: Cole-Parmer U80080-10
- Electronic oxygen monitor:
  - (a) small: GC Instruments monitor
  - (b) sensor plus external monitor: Teledyne 320B/RC (with remote sensor) or Teledyne 3190 (trace analyzer with remote sensor)
  - (c) hand-held head-space analyzer: PBI-Dansensor Checkpoint; movable head-space analyzer with printer: PBI-Dansensor Checkmate
- Datalogger for RH, temperature, and oxygen monitors: Conservation Support

**For humidification system:**

- See list of supplies after description in Appendix I.

**For leak testing:**

- See description of halogen leak detector in Appendix I.

**For nitrogen supply:**

- Liquid nitrogen (with heater, internal or external): see list at end of Protocol A
- Nitrogen cylinders (mounted in a cylinder bank or on cradle and attached to manifold: see list at end of Protocol A
- Nitrogen generator (membrane type): Air Products, Permea Division, PRISM®, Models 1100 or 1300

**For low-oxygen monitors (alarms set for 19.5% oxygen or less):**

- Room oxygen monitor with alarm, mounted at head height near enclosure: GC Industries
- Personal oxygen monitor with alarm: GC Industries

**For nitrogen removal (alternative method):**

- Vacuum cleaner, tank type, local household type
- Swagelok fitting to connect nitrogen exit port to vacuum cleaner hose: B-400-R-12 or B-600-R-12

## Protocol D: Anoxia in Large Rigid Chambers

Unlike the preceding protocols, which contained detailed instructions for large, user-made or commercial, plastic-film enclosures, Protocol D is a general discussion of using a large rigid chamber for insect eradication. Typically these commercial chambers are constructed of heavy steel plates, although they can also be made from thick, rigid plastic panels.

Although initially costly, a rigid chamber is particularly suited to an institution with objects, such as those in ethnographic collections, that have a continuing insect infestation problem. In the long run, a fully automated rigid chamber that is a permanent fixture would be cost-effective. The recommended size range for these chambers is 3 $m^3$ to 7 $m^3$.

Rigid chambers are purchased fully equipped by the manufacturer as a turnkey package. The only parts that must be supplied are shelving or similar accessories for placing museum objects inside the treatment chamber, and the nitrogen source. These units must be purchased from an experienced and competent manufacturer and made to the conservator's specifications. Any attempt to save money by having a chamber made by a local fabricator risks expensive repair and modifications. See "Rigid Chambers for Anoxia" in Chapter 4 for additional information.

Although a rigid chamber requires no especially large working area or high ceiling, its installation must be planned carefully, for example, to avoid narrow hallways when moving the chamber from the loading dock to its final installation site. The choice of nitrogen source—liquid nitrogen or a nitrogen generator—is a factor in the room design. (A nitrogen generator is more expensive but far less labor intensive, and it is the best match to an automated chamber's operation.) The only other consideration for a permanent, rigid chamber is how to exhaust the waste nitrogen after treatment. A room with direct access to the outside is preferable, but a direct duct to the outside is certainly adequate.

The following describes an anoxia treatment in a fully automatic 3 $m^3$ steel chamber that is integrated with a nitrogen generator and a vacuum pump. The chamber's leak rate was established during use at GCI at 50 ppm of oxygen (150 cc) per day, the maximum allowable leak rate. Several dozen small items requiring an RH of approximately 55% are to be treated.

### Preparation: Testing for leaks

The chamber's door is opened slightly, halogen gas is injected inside, and the door is quickly closed. Then the nitrogen generator is activated to release a slow flow of 99.99% pure nitrogen into the chamber. Then the chamber is set to automatically pressurize to a safe value of 3 barg (5 psig), after which all valves to the chamber are closed and nitrogen production is stopped.

The halogen leak detector is moved over the chamber's exterior edges where two surfaces meet and over areas where connectors and ports penetrate the unit, with special attention paid to the areas around the door and all connections. Any leaks discovered must be repaired. This may involve repairing or replacing gaskets, reseating connectors to O-rings and gaskets, increasing the pressure on O-rings and gaskets, or plugging holes with an epoxy filler.

After leak testing is completed, the control system is set to flush the chamber with air and exhaust the contents to the outside of the building. The control system turns off when the chamber's oxygen content is a safe 20%, at which time the chamber door can be opened.

At this point metal shelving can be added to the chamber and the objects to be treated positioned on the shelves. Then the door is closed and tightly sealed with its pressure latches.

**Treatment:** For the treatment example used here, the controllers on the system's control panel are set to an RH of 55%, an oxygen level of 0.1%, a temperature of 25°C, and a duration of fourteen days. Alarms are set to go off, and the system to shut down, if conditions inside the chamber deviate by greater than 5% for RH and if the oxygen level goes above 0.3% and the temperature drops below 23°C.

Switches are thrown to activate the nitrogen generator, recorders, and monitors. At a set interval, the control system outputs treatment conditions to both memory and to a monitor screen. At the end of fourteen days, the switch to change to "Air Flush Mode" is thrown, forcing air into the chamber as the nitrogen is evacuated to the outside of the building.

When the oxygen level reaches 20%, the unit shuts down. The door can now be opened and the objects returned to galleries or to storage.

## Materials and equipment

Except for shelving, equipment and all connections and instrumentation are supplied by the manufacturer as a turnkey package. See Appendix I for manufacturers and suppliers.

**For rigid chambers:**
- Manufacturers include Vacudyne, Thermotron, Bemis

**For nitrogen supply:**
- Liquid nitrogen (with heater, internal or external), or
- Nitrogen cylinders (6 mounted in a cylinder bank, or six pack, or on a cradle) and attached to manifold that leads to the chamber; or
- Nitrogen generator (membrane type): Air Products, Permea Division, PRISM®, Models 1100 or 1300
- Nitrogen two-stage regulator, with gauges: Cole-Parmer, U-98202-23
- Nylon tubing, copper tubing, and necessary connections from the nitrogen supply to the chamber (available from the chamber manufacturer)

**For nitrogen humidification:**
- Fully automated humidification units are available from the chamber manufacturer.

**For leak testing:**
- Halogen leak detector: see Appendix I.
- Ultrasonic leak detectors: American Gas (This is an alternative leak detector that can be used only with a rigid chamber. With an ultrasonic leak detector, a small ultrasonic generator or source is placed in the chamber, and the

detector measures changes in sound transmission to pick up leaks.)

**For low-oxygen monitors (alarms set for 19.5% oxygen or less):**

- Although an automated system automatically vents all nitrogen to the atmosphere outside the work area, it is a wise precaution to have one or more oxygen monitors in the treatment area. Personal oxygen monitors and monitors for mounting at head height near the chamber, all with an alarm, are available from GC Industries.

## Troubleshooting large-scale anoxia systems

Large-scale treatments require closer monitoring of the anoxic conditions than the same treatments in small bags. This is certainly true when a system has nitrogen flowing slowly through the enclosure throughout the treatment period; in this case, drastic changes in relative humidity or oxygen concentration can occur in eight hours or overnight. This troubleshooting section describes the steps to take if problems are experienced in controlling (1) the oxygen concentration and (2) the relative humidity or (3) if the nitrogen concentration near the enclosure increases enough to be a human safety hazard.

### 1. Oxygen concentration problem

*Symptom:*

Rapid oxygen increase inside the enclosure (i.e., any rate greater than a very gradual trend over several days) during an anoxia treatment.

*Possible Causes and Solutions:*

(a) For a system with continuous flow of nitrogen through the enclosure: If a continuous nitrogen flow was necessary, this means that the enclosure is relatively leaky. Thus the oxygen concentration will increase quickly if the nitrogen supply should stop. The most probable cause of the problem is exhaustion of a nitrogen cylinder or tank. Check the flowmeter. If the rate has slowed or stopped, check the cylinder pressure (or the contents of the tank, if liquid nitrogen is being used) to make sure that there is still adequate nitrogen available. If nitrogen is flowing normally, a radical new source of leakage has developed.

Run a leak check with a halogen leak detector after the following steps are taken:

- Stop the flow of nitrogen by closing the entry valve to the humidification system.
- Shut off the vacuum pump.
- Close the nitrogen outlet valve on the enclosure.
- Turn off the nitrogen source entry valve to humidification system.
- Open the relief valve adjacent to the inlet flowmeter of the anoxic enclosure.
- Close the nitrogen inlet valve to the enclosure and at the valve, detach the tubing from the flowmeter.

With the halogen leak detector ready, open the nitrogen inlet valve and inject some halogen gas into the enclosure, then shut the valve. Reattach the inlet flowmeter tubing to the inlet valve, then do the following:

- Restart the nitrogen flow from the source by opening the entry valve to the humidification system.
- Let many liters of nitrogen flow out the relief valve adjacent to the enclosure's inlet flowmeter.
- Open the enclosure's nitrogen inlet valve and close the relief valve.
- Let enough nitrogen flow in to the enclosure to cause it to swell slightly. In the case of a rigid chamber, let enough nitrogen flow in to cause a slight pressure increase (to 3 barg).
- Immediately open the relief valve and shut the nitrogen inlet valve to the enclosure.
- Close the nitrogen source entry valve to the humidification system.

Now inspect the entire enclosure for new leaks with the detector, starting with the obvious problem spots—connectors, seals around the base (or the door in a rigid chamber) and on top of an enclosure or Bubble. Seal or otherwise correct all leaks found as described in each protocol. Some small holes and tears in plastic film can be temporarily patched with self-adhesive aluminum tape or a small piece of the plastic film glued with rubber cement. This is easily done along some seals, but it is very difficult to do in a wall because the patch cannot be pressed firmly in place without an opposing pressure from the other side of the film.

*Note:* Liquid leak detectors, made for use with systems under relatively high pressure, might be useful for an enclosure that is only mildly pressurized. A leak from a pinhole or crack in such a system would produce tiny bubbles as gas slowly exits. One brand of liquid leak detector is called Snoop, available from Swagelok. Squirt a bit of Snoop on all questionable points, then wait 30 seconds or a minute to see if small foamlike bubbles form, indicating a leak.

Once leaks have been repaired, restart the treatment, beginning with the steps described in Protocol B, Part 2 ("Anoxic treatment"), and run for another fourteen days.

(b) For a system that showed no leaks initially and did not require a steady nitrogen flow to maintain a low-oxygen condition: Causes for a rapid increase in oxygen level include improper seal between enclosure base and cover, imperfections at the heat-sealed seams, and a hole or tear in the plastic film caused after the seams were sealed and checked.

The problem must be due to a small leak. Perform the leak test (described in Appendix I) by injecting a small amount of tracer gas into the enclosure and filling it with nitrogen until a slight positive pressure is established inside. Check for leaks especially around the bottom seal, heat-sealed seams, and connectors. Reseal any leaks and repeat the leak testing until the leak detector cannot find any new trouble spots. Small holes and tears can be temporarily closed by applying self-adhesive aluminum tape or a small piece of film glued with a rubber cement.

**2. Relative humidity problem**

*Symptom:*

Rapid or large change of RH inside the plastic-film enclosure or rigid chamber

*Possible Causes:*

- Faulty operation of the humidification system
- Change in temperature in the anoxia enclosure

*Solutions:*

Faulty operation of the humidification system: Check the flowmeter attached to the nitrogen supply. If the reading is not the same as at the start of treatment, increase or decrease the nitrogen flow until a desired flow rate has been achieved.

Check the RH of the nitrogen exiting the humidification system by first opening the relief valve adjacent to the enclosure's inlet flowmeter, closing the inlet valve, and detaching the tubing to the inlet valve. Then attach an RH calibration bag to the tubing as described in "Construction and Use of a Nitrogen Humidification System," Step 3 in Appendix I. Do *not* change the valve settings on the humidification flowmeters; continue to maintain the same overall flow rate that was in effect when the problem occurred. This assures that the RH reading from the sensor in the calibration bag will be the reading of the nitrogen being introduced into the enclosure when the difficulty was noted. Is the RH reading the same as the problematic RH recorded in the anoxia enclosure? If so, adjust the valve settings on the humidification system flowmeters to produce nitrogen with the desired RH.

Quite large RH variations are caused by temperature changes inside the enclosure, which in turn are due to a rise or fall in the ambient room temperature. A temperature drop from the desirable 25°C to 20°C, which dangerously extends the time required for killing insects, also increases the RH from 50% to 62%. Similarly, a rise in temperature from 25°C to 30°C reduces the RH from 50% to 35%, which would not be favorable for some wooden objects. To minimize large RH fluctuations, maintain room temperature as uniformly as possible at 25°C throughout the anoxic treatment period.

**3. Increased nitrogen level around enclosure exterior**

*Symptom:*

The low-oxygen monitor sounds an alarm indicating that less than 19.5% oxygen is present in the air near the anoxic enclosure.

*Possible Causes:*

- Oxygen concentration at the treatment site has actually dropped below 19.5%

- Malfunction of the low-oxygen monitor
- Poor ventilation in workroom where anoxia enclosure is located

*Solutions:*
If the low-oxygen alarm sounds as the enclosure is being raised after an anoxic treatment, nitrogen is flowing from the enclosure into the room. Exposure to a moderately oxygen deficient environment, such as 19.5% oxygen, for a few seconds is not fatal. There is enough time for workers to quickly turn on a mechanical room ventilator and/or open windows and doors to improve nitrogen dissipation; turn off a main nitrogen valve, *and then leave the room as quickly as possible.* It is not advisable for anyone to be exposed to the low-oxygen condition for more than a few seconds.

*Do not* assume that the oxygen monitor is sounding a false alarm. *Do not* reenter the treatment site unless equipped with a personal low-oxygen alarm or a self-contained breathing apparatus.

Wait until the low-oxygen alarm stops, indicating that the oxygen level is above 19.5%, before proceeding. It is essential to have two people, each with a personal oxygen monitor, reenter the room to look for the source of the problem. In this case, it is most likely a massive leak at a connection point with the enclosure, with the nitrogen tubing, or with the humidification system, but leaks can occur elsewhere. Using a liquid leak detector such as Snoop, check for leaks at all connection points: on the nitrogen cylinder or tank and with valves in the humidification system and with the treatment enclosure. Tighten any leaky connections.

Next make sure the tubing on the vacuum pump that exhausts the enclosure is unclogged, and check that the pump is working properly. In the unlikely event that the room's low-oxygen monitor is malfunctioning, this will become apparent if after an hour or so with maximum ventilation in the work area, the alarm has not stopped. Double-check the situation by going into the room with a personal oxygen monitor. If its alarm does not sound, then the room's low-oxygen monitor is defective and should be repaired.

## Notes

1. ESCAL™ is a silica- or siloxane-coated, 0.11 mm thick film with water vapor permeability of 0.01 g/m²/day and an oxygen transmission rate of only 0.05 cc/m²/day. Treatment bags made of ESCAL™ require a minimal quantity of oxygen absorber and RH-conditioned silica gel to maintain an optimal, constant RH and low-oxygen environment over the long term. ESCAL™ is sealed like any other oxygen-barrier film. It is available in flat sheets, 1 m wide by 100 m long. ESCAL™ is also sold as a flat tube in widths ranging from 12 cm to 27 cm, making the production of small pouches quick and easy: simply cut off the desired length of the ESCAL™ tube and seal the two ends when ready to begin treatment. Such a bag can maintain an anoxic atmosphere for about ten years, compared to one year for a bag made from a typical oxygen-barrier film.
2. Although heat sealing treatment bags and pouches is easily done with a simple tacking iron, the task can be onerous if one hundred or more bags need to be made at once. Fortunately, a variety of heat sealers are available to speed up the process.

Impulse sealers that quickly seal a 120 mm long seam are relatively inexpensive (approx. $175). Larger sealers that make a 320 mm seam are about $575; those with a 620 mm sealing bar are about twice that amount. (Although the small sealer is inexpensive, it means that multiple joins have to be made in a long edge; its cost saving is offset by the possibility of creating improper seals when many pouches must be prepared.) A foot pedal attachment for the sealer further simplifies the process and costs less than $100. With a foot pedal, the hands are free to stretch the two surfaces being sealed to avoid wrinkles in the resulting seam.

Appendix I

# Protocol Tools

This Appendix explains the tests, techniques, and calculations required by the anoxia protocols. It should be read before attempting treatment.

## Leak Testing of Anoxic Enclosures

Types of leak detectors are described in Chapter 2, "Monitoring of Anoxic Conditions." A halogen leak detector is the least expensive and most effective system. *Halogen* refers to the tetrafluororethane tracer gas, such as computer cleaners and refrigerants, typically used with this type of detector, which is described here and shown in Figure 2.13 (p. 39). The tracer gas is released inside the treatment enclosure, causing the thermal conductivity of the enclosure's atmosphere to change at leakage points. This change is picked up by the sensor on the detector nozzle, triggering an alarm.

The halogen leak detection procedure described here is for a small plastic treatment bag used in Protocol A, although it works well for all types and sizes of enclosures. Leak testing large enclosures is addressed in the appropriate protocol.

**Step 1:** Leak testing is done after the objects to be treated, oxygen absorber, and monitoring devices have been placed inside the treatment bag. Then heat seal the remaining seam, leaving a 1 cm to 2 cm opening in one corner of the bag. Following the detector manufacturer's recommendations, insert a small amount of the halogen tracer gas through this opening. Then quickly heat seal the opening. (For leak testing a large enclosure that has been fitted with nitrogen inlet and outlet ports with valves, as in Figure 5.18 (p. 107), simply open a valve, inject the tracer gas, then close the valve.)

Optional: Slightly pressurizing the bag with nitrogen increases the leak rate and improves detection. This is done immediately after inserting the tracer gas. Using the same opening, introduce nitrogen through a tube until the bag is *barely inflated*—overinflating the bag can open seams. Quickly remove the nitrogen tube and heat seal the opening.

**Step 2:** Allow a few minutes for the tracer gas to mix throughout the treatment bag. Then run the detector nozzle along the seams and corners (and around all connector ports in large enclosures) as shown in Figure I.1. This is where the film is stressed and where pinhole

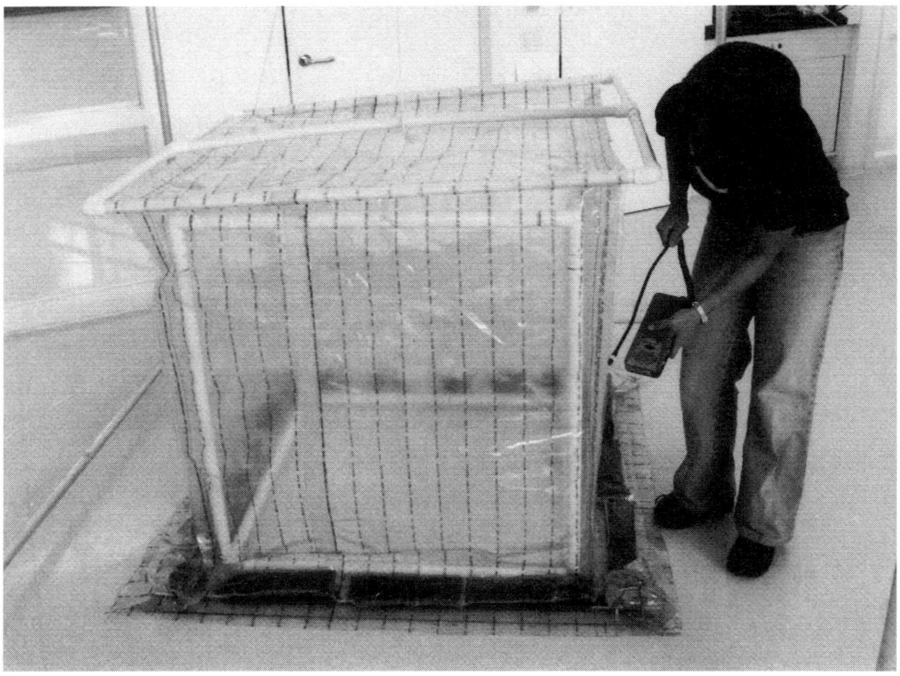

**Figure I.1**
A halogen leak detector is used to verify the airtightness of the enclosure.

leaks might occur. Leaks will trigger an alarm from the detector and, in some models, a flashing light.

**Step 3:** Reseal imperfect seams and corners with a tacking iron, although this is never effective over a flaw, such as a leaking wrinkle. If resealing does not work, bypass the leak with a second seal. Start at the seam just before the leak, seal toward the center of the bag along an arc about 1 cm from the leak point, and end back at the seam just after the leak. To repair a leak around a connector in a large plastic enclosure: (a) remove the connector, (b) close the hole by heat sealing a patch of film over it, (c) open a new hole in the enclosure, and (d) reattach the connector. Pinholes can be covered for the duration of an anoxia treatment with self-adhesive aluminum tape. After all flaws have been repaired, recheck for leaks with the detector, and reseal as needed.

### Supplies required for leak testing

See Appendix IV for supplier information.

- Halogen leak detector:
- Model 5750, TIF Instruments
- Model FGT-202, American Gas
- Can of halogen tracer gas (available at local refrigerant or computer dealers)
- *Optional:* nitrogen gas cylinder or other nitrogen source for pressurizing the bag. This should be fitted with a two-stage regulator and nylon or copper tubing for inserting into the anoxia bag. The dry nitrogen does not need to be humidified because of the relatively small volume of gas used and the short time that the objects will be exposed to the lowered relative humidity caused by the nitrogen.

## Construction and Use of a Nitrogen Humidification System

Large treatment enclosures are purged with pure nitrogen to create an anoxic environment. However, the dry nitrogen must first be humidified to avoid desiccating the objects being treated (see Chapter 4). Commercial, fully automatic humidification units that adjust the moisture content of dry nitrogen gas to any desired relative humidity are available, but they are expensive. Sources include Vacudyne, Power Plastics, and Kojima Seisakujo (see Appendix IV).

This section describes a simpler system that can be easily built. The system, designed by a conservation scientist, consists of two bottles. The first bottle, known as a bubbler or nebulizer, contains water through which a stream of nitrogen is bubbled to yield a nearly saturated gas flow. The humidified gas then passes to a second, dry bottle, known as a droplet collector. Here water droplets are removed from the wet gas and collected in the bottle. A measured rate of nearly saturated gas is mixed with a measured rate of dry nitrogen to produce a moisture content at the desired level. Such a unit can also be purchased ready-made from Metode in Spain (see Appendix IV).

Figure 4.10 (p. 68) is a schematic diagram of the humidification system in operational mode. Figure I.2 shows the system in calibration

**Figure I.2**
The humidification system in calibration mode, with RH monitor in plastic bag

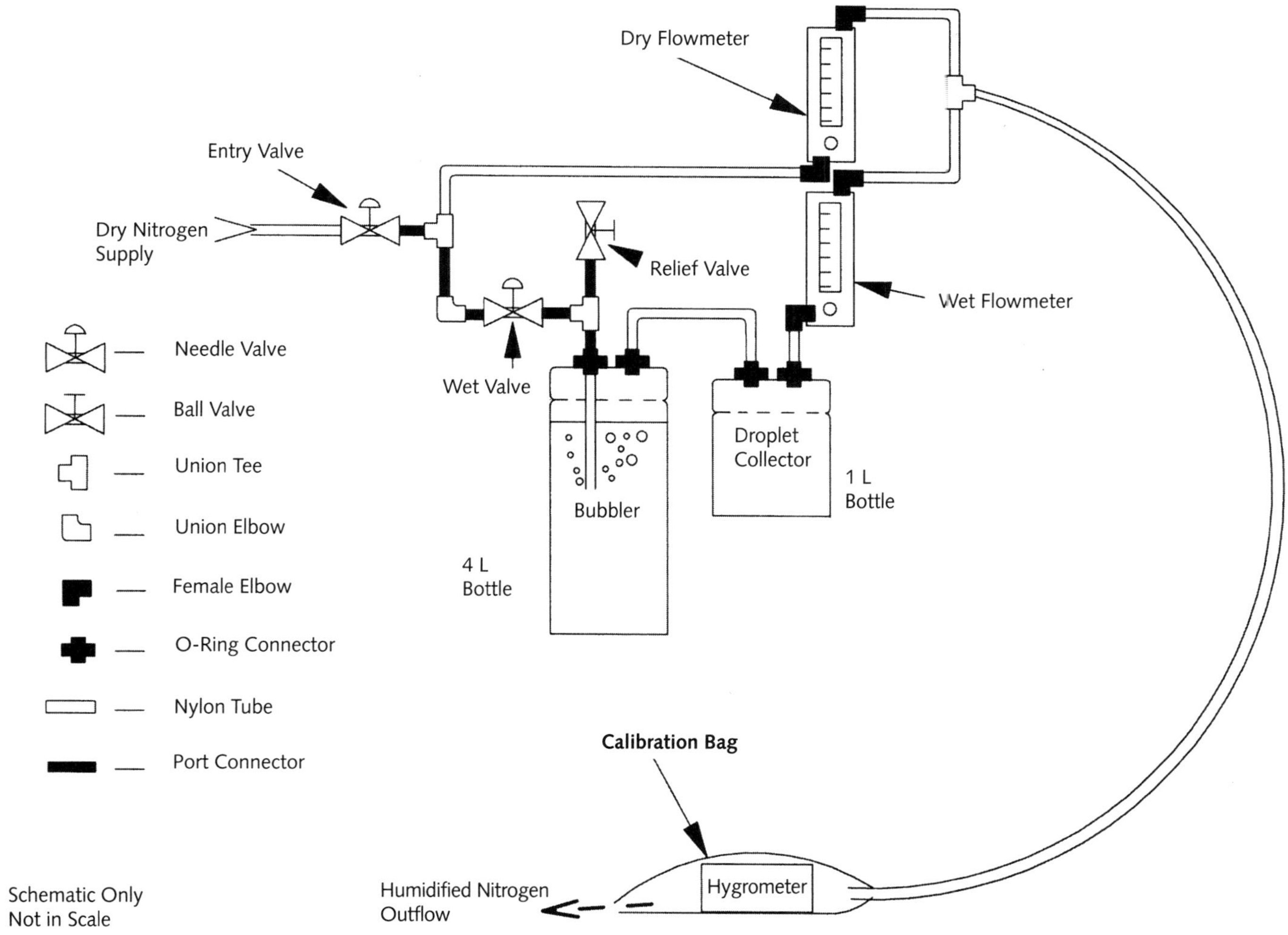

mode with an RH sensor attached to the system's outflow. Figure 4.11 (p. 69) shows the types of valves and connectors used in the system, and Figure 4.12 (p. 69) is a photograph of the humidification system.

Specific supplies and part numbers for the materials needed to fabricate this humidification system are given at the end of this section. The essential components are

- 1 4 L and 1 1 L polypropylene or HDPE bottle
- 2 gas flowmeters
- 2 Whitey needle valves and 1 Whitey ball valve
- 4 Swagelok O-ring connectors (plus appropriate washers and nuts)
- 3 Swagelok tees
- 2 elbows
- 6 port connectors
- Nylon laboratory tubing
- Fast-reacting (capacitive) hygrometer to calibrate the RH of the humidified nitrogen

**Tools required:**
- adjustable wrenches
- power drill with 7/16 in. and 9/16 in. brad point drill bits
- 1/4 in. drill bit for metal

### Constructing the humidification system

**Step 1 (drill the bottle caps):** Each bottle cap will have two holes fitted with O-ring connectors to hold tubing in place. The O-ring in the connector makes contact with the top of the cap and is held tightly to it by a washer and nut on the underside of the cap (the nuts should be brass rather than steel to avoid corrosion). The threaded end of the connector extends through the hole. The attachment of an O-ring connector to a bottle cap is as shown in Figure I.3.

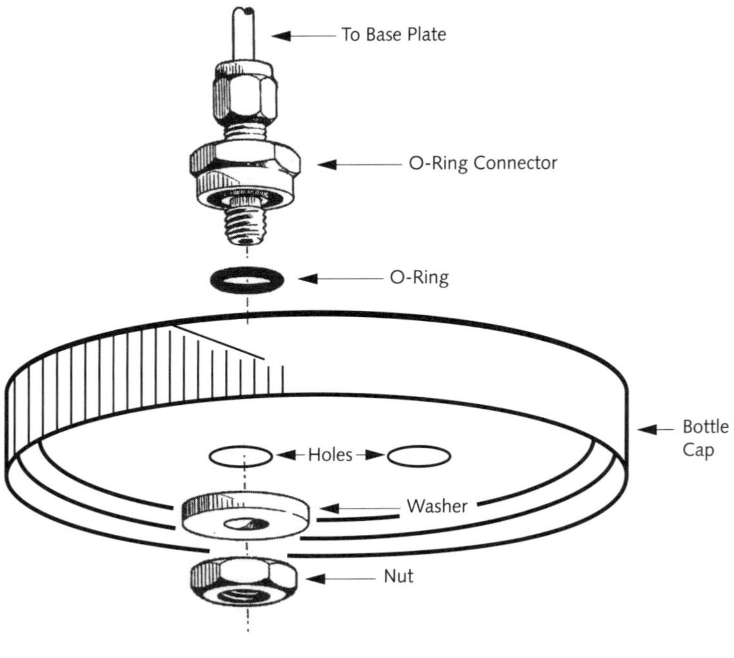


**Figure I.3**
O-ring connectors are installed on a bottle cap of the humidification system.

Start by drilling the two holes in each cap, starting from the top, using a $^{7}/_{16}$ in. brad point drill that matches the threaded end of the O-ring connector. The holes must be far enough apart so that the O-ring connectors and their retaining nuts and washers do not interfere with each other. Carefully sand the tops of the caps around each of the holes to remove any ridge left from drilling. The area around the holes must be smooth to insure a perfect seal with the O-ring.

**Step 2 (prepare 4 L bubbler water bottle):** First enlarge the hole in one of the O-ring connectors by drilling a $^{1}/_{4}$ in. diameter bore $^{1}/_{3}$ in. deep, starting from the end of the connector that will extend above the cap. Then force $^{1}/_{4}$ in. nylon tubing into the bored hole from the same end. Cut the tubing so that it will end 2 cm to 4 cm from the bottom of the bottle and extend $^{1}/_{3}$ in. above the cap. Then drill two $^{1}/_{16}$ in. holes completely through the tubing that extends into the bottle, placing the holes 2 cm to 4 cm above the end. This will allow nitrogen to bubble out of the tubing and up through the water in the bottle.

Next, grease all four O-rings in the connectors lightly with vacuum grease. Then insert each O-ring connector into the appropriate hole in the bottle caps, tightening each connector firmly in place with a nut and washer on the underside. Refer to the diagram of the completed system in Figure 4.10 (p. 68).

**Step 3 (fabricate RH calibration bag):** Before the humidification system is put into operation, the RH of the nitrogen exiting the unit must be calibrated. This is done by passing the nitrogen stream through a length of nylon tubing into an open plastic bag containing an RH sensor (hygrometer). Figure I.2 (p. 127) shows this arrangement. The bag and tubing are removed after the humidification unit is calibrated.

To make the RH calibration bag, use a 15 cm × 40 cm piece of plastic film to make a bag with a 3 L to 4 L volume. If you have only a tacking iron or bar sealer, use any oxygen-barrier film. The bag can be made of polyethylene film, but this film is difficult to seal without an impulse sealer. Fold the piece of film in half to measure 15 cm × 20 cm, then heat seal the two 20 cm sides.

Cut a corner of the closed end of the bag to make an opening just large enough to insert about 15 cm of a 1 m length of nylon tubing. Securely tape the corner to the tubing with any variety of tape. (This bag will not be a permanent fixture; the taping need only be good enough to hold the bag tightly to the tubing.) The other end of the tubing will be connected to the outflow of the completed humidification system for calibration.

Insert an RH sensor into the plastic bag through the unsealed edge and place the unit on a table or support so that the humidified nitrogen will be able to flow freely over the RH sensor with its display clearly visible. Then, using tape, loosely gather the open end of the bag into an opening roughly 8 cm to 10 cm in diameter. This will allow unrestricted exit of the nitrogen gas.

**Step 4 (assemble system components):** Using the photograph in Figure 4.12 (p. 69) and the schematic in Figure I.2 (p. 127) as guides, connect the tubing, tees, elbows, and other fittings, O-ring connectors, valves, flowmeters, and RH calibration bag.

**Step 5 (build mounting):** Mounting the completed humidification unit provides stability and portability. A mount can be made with standard chemical laboratory supports and clamps or by fabricating a metal base and frame with appropriate holders for the bottles and components. A plywood mounting is an easy alternative, and its construction is described here (see schematic in Fig. I.4).

Start with a vertical piece of plywood attached to a base, strengthened by right-angle shelf supports or their equivalent. Cut out two rectangular holes in the vertical board into which the bottle will be inserted. The cutout should be 20 cm to 30 cm higher than the larger bottle and 60 cm to 80 cm wider than the two bottles together. Position the bottles in this cutout such that the tubing leading from the bottles can be neatly attached to the board, along with the flowmeters, via nylon ties. The bottles should set on a 5 cm thick piece of board that is removable. When the wood is removed, the large bottle can be unscrewed and lowered for refilling with water when necessary.

Temporarily position the completed assembly on the wooden mount and mark where holes should be drilled for nylon ties to support the various components. Remove the components, drill the premarked holes, reassemble, and secure the components to the board with nylon ties. Wedge small pieces of wood (shims) behind components so that all have a uniform distance from the mount and there is enough room to turn valves.

**Step 6 (final assembly):** The following instructions assume that the completed humidification system is mounted, as described above.

(a) Remove the container portion of both bottles from their caps. Then wrap the threads of both containers clockwise

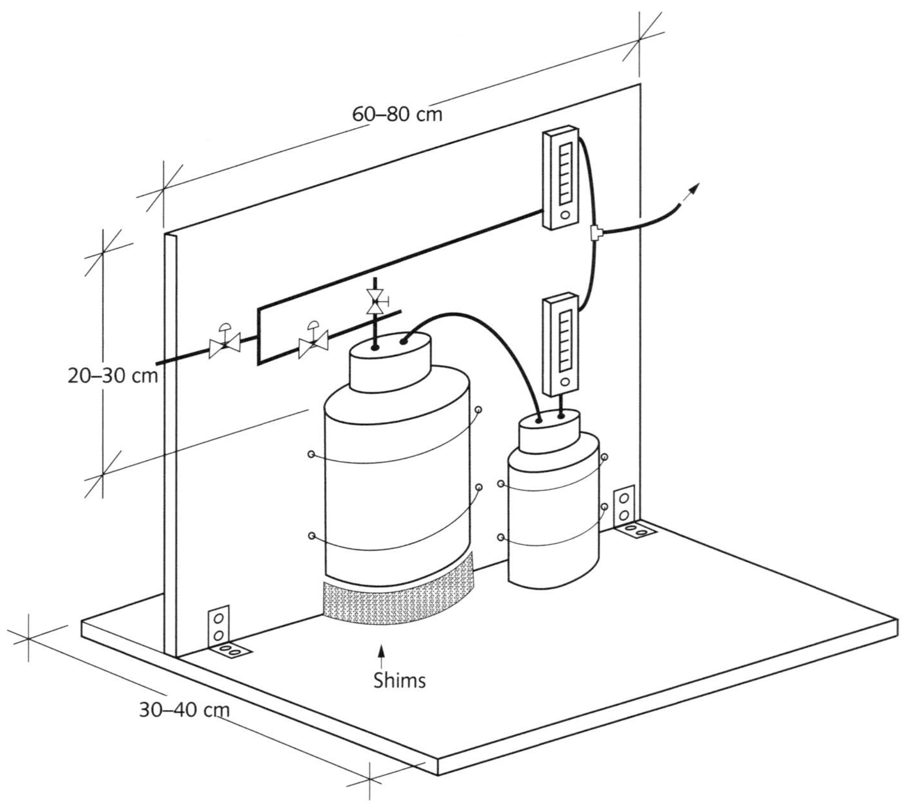


**Figure I.4**

Schematic of a plywood mounting of the humidification unit

with Teflon pipe-thread tape. Fill the 4 L container two-thirds full with distilled water, then screw both containers back onto their caps.

(b) Using copper or nylon tubing and a Swagelok fitting, connect the nitrogen cylinder (or other nitrogen source) to the needle valve at the entry to the humidification system.

(c) Attach the RH calibration bag made in Step 3.

**Step 7 (system check):**

(a) Check to see that the entry valve between the nitrogen source and humidification unit is closed and that the exit from the RH calibration bag is unblocked.

(b) Referring to Figure I.2 (p. 127), completely open the valves of both the dry flowmeter and the wet flowmeter. With two full turns until they are almost fully open, open the relief valve on top of the tubing in the 4 L water bottle (bubbler) and the wet valve that controls the amount of nitrogen entering the water bottle. The rate at which the nitrogen will bubble through the water can be increased by slightly closing the valve on the dry flowmeter.

(c) With the entry valve to the humidification unit still closed, open the main valve of the nitrogen cylinder and turn the diaphragm valve on its regulator clockwise until the gauge indicates about 1.4 to 2.8 barg (20–40 psig).

(d) Check the nitrogen cylinder for leaks with a liquid bubbling agent such as Snoop: put a few drops on every connection to and from the regulator and to the humidification unit's entry valve. The appearance of foam will reveal the location of any leaks. Tighten leaking connections.

(e) Open the entry valve slightly, just to get a sense of how the nitrogen will flow through the system. Then open it completely. Next, close the relief valve on the bubbler bottle and, using the wet valve, gradually increase the nitrogen flow until the readings on the two flowmeters add up to one-tenth the volume of the treatment bag/enclosure per minute.

(f) Adjust the wet valve and the dry-flowmeter valve until the rates on the wet and dry flowmeters are about equal; this will produce a 50% RH nitrogen flow. It is best to start with both the wet valve and the dry-flowmeter valve nearly completely open and then slowly close the dry-flowmeter valve to increase the bubbling and resulting flow of humidified nitrogen.

(g) Check all joints and both bottle caps for leaks with Snoop or other bubble-forming leak detector. Tighten any leaking connections.

**Step 8 (system calibration):** The calibration of the humidification unit to achieve 50% RH (or another preferred RH level) should be made at several different flow rates between two extremes in order to understand the range of the system's capability. Arbitrarily, a maximum rate might be 30 L/min, and a minimum rate might be 5 L/min.

When the flow registered by both the wet and dry flowmeters is close to one-tenth the volume of the treatment enclosure per minute, record the RH reading on the sensor in the calibration bag. Obtaining the correct flow rate may require opening the entry valve further. If this does not increase the flow enough, the diaphragm valve on the nitrogen regulator should be turned clockwise a little more to increase the nitrogen pressure and thus flow.

Repeat this process of finding what valve settings produce 50% RH for three or four different flow rates.

When a stable RH is obtained, record the position of the open wet valve and the open dry-flowmeter valve as accurately as possible for future reference. A suggested method is to record the position of the white index mark on the valves in terms of a clock: 9 o'clock for a quarter-turn open, 6 o'clock for a half-turn open, and so on.

**Step 9 (system shut down):** Shut down the system as follows: (a) Open the relief valve on the water bottle and then turn the diaphragm valve on the nitrogen cylinder regulator counterclockwise until the nitrogen flow stops. (b) Close the entry valve so that there is no pressure in the line to the rest of the humidification system.

**Warning!** This shut-off procedure is to be used only after performing a *calibration*. The shut-off procedure when the humidification system is *in operation* and attached to a treatment enclosure is different. In that case, begin the shut-off procedure by first opening the relief valve near the treatment enclosure and immediately closing the inlet valve to the enclosure. Then continue with the sequence described in Step 9.

## Supplies and specific part numbers for materials required to fabricate a nitrogen humidification unit

See Appendix IV for supplier information.

**Available from Cole-Parmer (2001–2002 catalog numbers):**

- 1 4 L polypropylene bottle, U-06057-20
- 1 1 L polypropylene bottle, U-06048-40
- 2 flowmeters with valves (4–40 lpm), U-03216-38
- Nylon tubing, 50 ft, U-06489-06
- Hygrometer (10 sec response time), U-37100-15

**Available from the Swagelok Company (2001 catalog numbers):**[1]

- 4 O-ring straight thread male connectors, brass, B-400-1-OR (1/4 in.)
- 3 union tees, brass, B-400-3 (1/4 in.)
- 4 female elbows (for flowmeters), brass, B-400-8-2 (1/8 in. NPT, 1/4 in. tube)
- 2 union elbows, brass, B-400-9 (1/4 in.)
- 6 port connectors, brass, B-401-PC (1/4 in.)
- 2 Whitey needle valves, brass, B-1RS4
- 1 Whitey ball valve, brass, B-42S4 (1/4 in.)

**Hardware items:**

- 2 adjustable small wrenches (6 in. or 150 mm), or a set of end wrenches
- Small electric drill (3/8 in. or 10 mm) with 1/4 in. drill bit for metal
- Drill bit set, brad point (for minimal damage to surface being drilled)
- 4 brass nuts, 7/16 in. × 20 thread (for tightening O-ring connectors)
- 4 brass washers, maximum 1/2 in. hole
- Assorted lengths of nylon ties
- Plywood for mounting humidification unit components. Alternative for mounting: use support rods, base, and clamps, or a fabricated metal mount.
- Duct tape or masking tape

**For plastic bag to hold RH sensor:**

- Any type of plastic film, 15 cm × 40 cm
- Appropriate heat sealer for the plastic film

## Leak-proof Connectors in Anoxic Enclosure Walls

Installing connectors to a large treatment enclosure, either for nitrogen supply and exhaust or to introduce electrical wires, is a simple process. It consists only of cutting a small circular hole in the oxygen-barrier film of the enclosure, inserting the appropriate connector unit, and tightening it to the film by screwing a washer and nut from the other side of the film (Fig. I.5). Holes can be precisely cut in the film with a sharpened laboratory cork borer opposed by a piece of corrugated cardboard on the opposite side of the film.

Connectors can be the source of leaks if not attached correctly. The difficulty lies in creating a tight seal where the barrier film is breached with the connector but not stretching or stressing the film at that point. To ensure a leak-free connection, lubricate the O-ring and the opposing washer with a small amount of an inert grease (preferably a vacuum grease such as Apezon Grease, Fomblin Fexfluorinated Grease, and DuPont Krytox Fluorinated Grease LVP), and rigorously avoid creating wrinkles or strains in the film when slowly tightening the nut and washer against the O-ring in the connector.

Because connectors are relatively heavy, it may be necessary to reinforce their point of attachment before cutting out the hole. This is done by heat sealing a small patch of the same barrier film to the inside of the enclosure wall at the attachment point.

In addition, as soon as the connector is attached to the enclosure, it must be supported so that

**Figure I.5**

A washer and a nut are used to press the plastic film against the O-ring to ensure an hermetic seal for the nitrogen inlet and outlet ports.

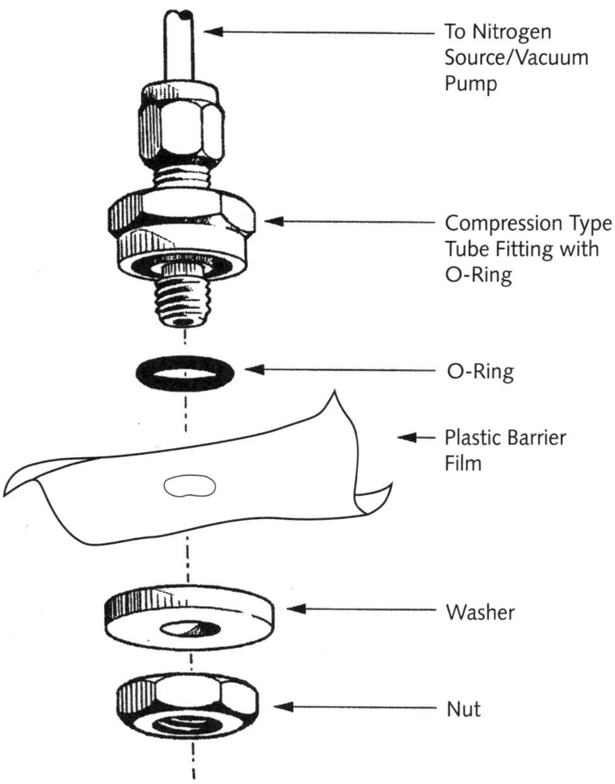

its weight and the weight of wires or nitrogen gas lines feeding into it do not strain the enclosure wall. (Possible supports are a chemical laboratory ring stand and clamp and a vertical board on a base with a hole for the connector, or attachment to a part of the enclosure's support frame. The electrical wires or nitrogen feed to the connector should also be protected from accidental movement that would put radical strain at the connector-film interface.

The following discussion outlines how to attach connectors to a plastic-film treatment enclosure.

**Electrical connectors:** Experience at the GCI has shown that most commercial electrical connectors are unsuitable for anoxic treatment enclosures because they are designed to be attached to rigid structures (see, e.g., the Web site of Pave, whose URL is listed in Appendix IV.) Either these connectors are unusually heavy or their flanges are not wide enough to hold an O-ring or gasket, which is essential for an airtight seal.

For anoxia treatment, it is advantageous to have a machinist fabricate the electrical connector from a rigid plastic material, for example polypropylene, to reduce the weight and avoid sharp edges. Figure I.6 shows the essential parts of an electrical connector suitable for use on a treatment enclosure: a threaded connector with an O-ring that will be pressed against the enclosure film; a washer for the opposite side of the film; and a nut that tightens both connector and washer together to form a seal with the film.

The wires are inserted through the connector and held in place with epoxy resin to make a hermetic seal. Figure I.7 shows how a clip holder is used to support the connector and inserted wires while the epoxy resin is added. Color-coded wires are advisable to avoid errors when connecting sensors inside the enclosure with their monitors on the outside.

Before attaching the connector parts to the hole in the anoxic enclosure (Fig. I.8), lubricate the O-ring and the opposing washer with a small amount of vacuum grease. Since

**Figure I.6, below**
Sensor wires and polypropylene parts used to make an electrical O-ring connector (left to right: screw threads, washer, threaded nut). (Scale = 5 cm)

**Figure I.7, bottom left**
Wires are sealed in the polypropylene electrical connector with epoxy resin.

**Figure I.8, bottom right**
Installation of an hermetically sealed electrical connector

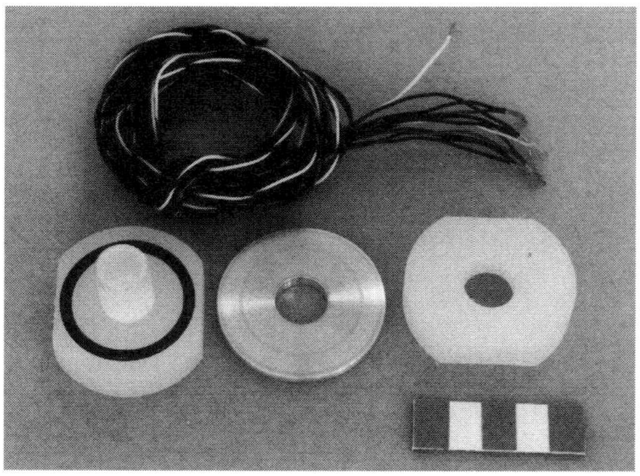

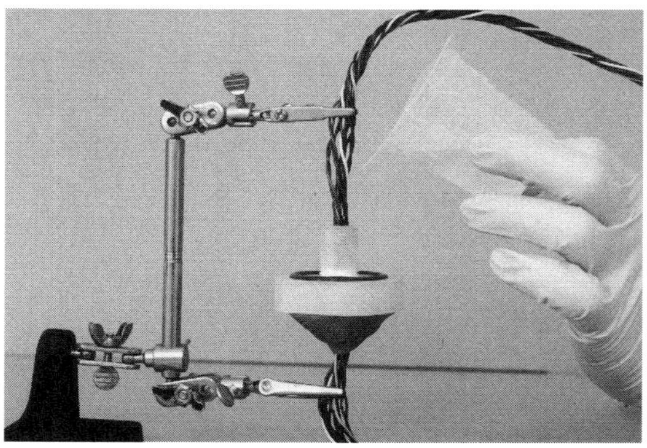

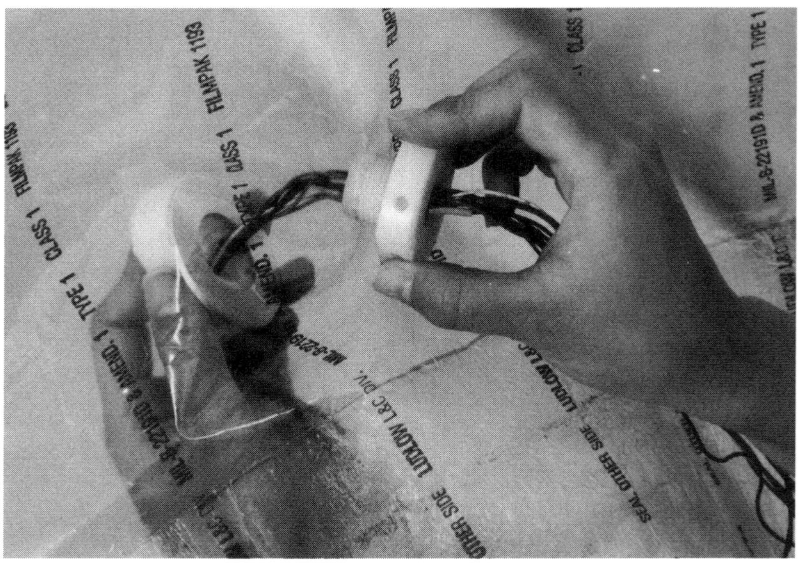

the washer is turned tightly on the threads of the connector, this lubrication greatly aids in avoiding any twisting or wrinkling in the film that would lead to leaks.

**Nitrogen inlet and outlet connectors:** Commercial O-ring connectors of brass or stainless steel, such as those made by Swagelok, can be used as the nitrogen inlet and outlet connections to a treatment enclosure. Figure I.5 is a schematic showing how such a connector attaches to the plastic film of a treatment enclosure using a washer and metal nut.

These precision connectors have the disadvantage that their O-rings do not rise much above the metal surface. Consequently, if the connector is overtightened, the metal rim can cut into the plastic film. Care must be taken to tighten these connectors as firmly as possible but not so much as to cause such damage. Connectors for nitrogen supply and exhaust should be located on opposite sides of the enclosure and at opposite heights to ensure good mixing of the nitrogen with the atmosphere and to prevent stratification (see Fig. 4.1, p. 59). Again, it is essential that the connector and nitrogen lines be supported and guarded from any accidental movement that would put undue strain on the connector-film interface and cause leaks.

### Supplies required for connectors

See Appendix IV for supplier information.

- Polypropylene rod, 2.0 in McMaster-Carr 8658K21 from which a machinist can make parts for an electrical connector
- 1.5 in. O-ring, electrical wires to sensors and instruments, slow-curing epoxy cement
- Swagelok O-ring male connectors, B-400-1-OR, washers with holes no larger than ½ in., and 7/16 in. × 20 nuts

## Calculation of Required Amount of Oxygen Absorber

*Important note: These calculations are only for the conventional oxygen absorber consisting of iron powder (Ageless®, FreshPax™, Atco™) and not for the RP type of absorber.*

How much oxygen absorber to use for insect eradication in small to medium-size bags or pouches depends on the volume of the bag and its gas-tightness. The volume can be roughly (and very adequately) estimated by multiplying the length and width of the bag by what its height will be after it is filled. The height is measured after draping the plastic to be used to make the bag over the object(s) as described under Protocol A (Fig. I.9).

The formula for calculating the amount of oxygen absorber required is

Amt. of oxygen absorber (in 1000 cc packets of oxygen absorbing capacity) = treatment bag's length (m) × width (m) × height (m) × 210

If the measurements of the treatment bag are in **cm** (centimeters) instead of **m** (meters), the conversion factor is **0.00021** instead of **210** in the above equation.

> Example: If a bag measures 1 m long and 0.5 m wide, and if the height of the filled bag is 0.2 m, then 1 m × 0.5 m × 0.2 m × 210 = 21.

Thus, 21 packets of Ageless® oxygen absorber with a reaction capacity of 1000 cc per packet (or 11 packets with a 2000 cc capacity) would be adequate to absorb the oxygen in that bag. This amount also would maintain the oxygen level near 0.1% for the 14 days of anoxia treatment in a *well-sealed* bag.

If a rectangular treatment enclosure is used, and its height is equal to the maximum height of an irregularly shape object to be treated, the amount of oxygen in the enclosure will be overestimated and, likewise, the amount of oxygen absorber needed. This overestimation will shorten the time to achieve anoxic conditions, and the ample supply of oxygen absorber will cope with normal leakage through the sides of the enclosure even after a low oxygen level is reached. (See Chapter 3, "Amount of Oxygen Absorber to Use for Anoxia in Bags," for a more detailed discussion of oxygen leakage and oxygen absorber behavior.)

It might seem economical to use less oxygen absorber than needed in an enclosure if objects occupy much of that volume, for example, densely packed books. However, the air content of objects is not trivial, and even most "solid" wood is about 30% to 40% air. The savings in the cost of oxygen absorber may not be worth the time spent making the complex calculations to figure the objects' volume in order to reduce the amount of oxygen absorber correspondingly. And there is the possibility of using too little of the essential oxygen absorber and not achieving full eradication.

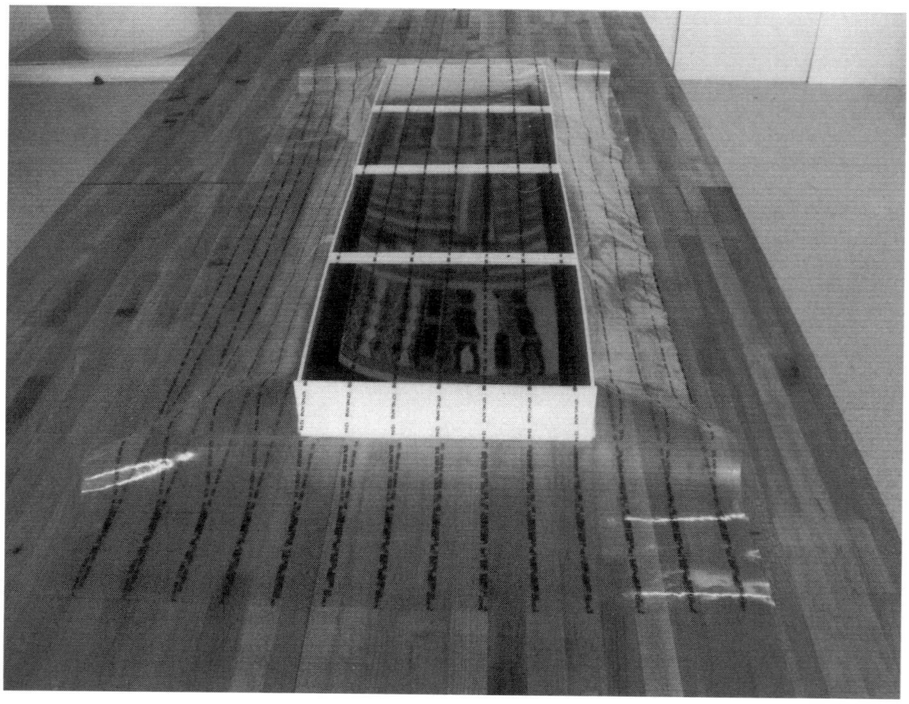

**Figure I.9**
The plastic film is folded loosely over the object to create a pouch (bag).

## Calculation of Required Amount of RH-conditioned Silica Gel

Moisture is released from the oxygen absorber used to create an anoxic atmosphere in small to medium-size bags. One Ageless® Z-1000 packet, for example, releases 0.4 g of moisture over time. Bead-type silica gel, conditioned to the most desirable RH for the objects to be treated, can fully compensate for this kind of water release. For example, each 100 g of bead-type silica gel conditioned at 50% RH will react with 2 g of water before it reaches equilibrium with an atmosphere at 55% RH. (The silica gel is conditioned by exposing 73.5 g of dry bead-type silica gel to an RH of 50% for at least several days to arrive at 100 g of conditioned gel.)

The amount of 50% RH-conditioned silica gel needed to maintain a 50% to 55% RH in an anoxia bag is calculated as follows:

Amt. of RH-conditioned silica gel =
Amt. of oxygen absorber (in 1000 cc packets) $\div$ 5 $\times$ 100 g

The treatment bag used as an example in the previous section required 21 packets of Ageless® with 1000 cc capacity each. Using that information in the formula above (21 $\div$ 5 $\times$ 100 g) yields 420 g as the minimum amount of RH-conditioned silica gel needed to compensate for the water released from the 21 packets of absorber.

### Materials

See Appendix IV for supplier information.

- Bead-type silica gel, Conservation Support (Silica gel in various sizes of packets are available from IMPAK and Royco. When conditioned to the desired RH, these packets are most convenient for small bags requiring only a few grams of silica gel rather than hundreds of grams.)
- Shallow plastic boxes, about 5 cm high, in various sizes to hold the silica gel in the treatment bag

## Notes

1. Swagelok O-ring connectors are given here in U.S. dimensions because of the large number of different threads used internationally. Swagelok has dealers all over the world, and these connectors can be obtained in metric or U.S. dimensions. Drill bits, nuts, and tubing in U.S. dimensions for the Swagelok fittings can be obtained on the Internet from McMaster-Carr; tubing, from Cole-Parmer. See Appendix IV for contact information.

# Appendix II

## Technical Addenda

### Reaction of an Iron Powder "Oxygen Absorber" with Oxygen

It is common knowledge that iron rusts because it reacts with oxygen. However, ordinary iron objects that are in relatively dry air react very slowly with oxygen. Oxidation is enormously accelerated when a piece of iron is ground into a fine powder because much more surface area is then available to react. It is accelerated even more when all or part of the powder is partly oxidized, changing the elemental iron, Fe, into $Fe^{II}$ oxides that themselves can rapidly be attacked by oxygen to make $Fe^{III}$ oxide. Additional accelerants are the presence of salt and water. This is because the difficult step in oxidation involves a transfer of electrons from the $Fe^{II}$ oxide to oxygen in forming $Fe^{III}$ oxide. Therefore, any ionic solution such as saline water that readily conducts electrons greatly expedites oxidation.

Oxygen absorbers consist of iron powder and $iron^{II}$ oxide with salt water. Expressed in the form of a chemical reaction, the process is

$$4\ Fe + O_2 + 4\ H_2O \rightarrow 4\ Fe(OH)_2 \quad \text{or} \quad FeO \bullet H_2O$$

Further oxidation of the $Fe^{II}$ to $Fe^{III}$ takes place to form a hydrated oxide rather than the never-isolated $Fe(OH)_3$, along with a large quantity of heat:

$$4\ Fe(OH)_2 + O_2 + 2\ H_2O \rightarrow [4\ Fe(OH)_3] \quad \text{or} \quad 4\ FeO(OH) \bullet H_2O$$
$$\text{or} \quad 2\ Fe_2O_3 \bullet 3\ H_2O$$

The final product of hydrated oxides is black, as was the original finely divided iron, in contrast to the red iron oxide so well known as rust.

### Calculation of the Amount of Oxygen Absorber Required to Maintain 0.1% Oxygen in an Anoxia Enclosure Whose Leak Rate Is Known

The rate at which an oxygen absorber reacts with oxygen leaking into an anoxia treatment enclosure must equal the rate at which the oxygen leaks in—if the oxygen level in the bag or container is to be unchanged.

> The amount of oxygen entering the well-sealed 100 L Aclar bag described in Chapter 3 (pp. 49–51) equals the 10.6 cc leaking through

the film, or about 11 cc per day. Calculations of this type are usually expressed in parts (e.g., of oxygen) per million of the atmosphere in the bag. Converting 11 cc in a 100,000 cc bag to ppm in the bag simply involves dividing 11 by 100,000 and then multiplying by a million, i.e., 110 ppm per day.

The relationship between the leak rate and the reactivity of the oxygen absorber is this (Lambert, Daniel, and Preusser 1992):

If Le = Leak rate of oxygen into the bag/day, then

$$\text{Ageless}^{®} \text{ capacity needed (cc)} = \frac{80 \times \text{Le (in ppm)} \times \text{Anoxia bag volume (in liters)}}{\text{Desired oxygen conc. (in ppm)}}$$

Thus with an Aclar bag of 100 liters and an oxygen level of 0.1% (1000 ppm), the Ageless® capacity required is 80 × 110 ppm × 100 L/1000 ppm or 880 cc. In other words, after the oxygen absorber needed for initial removal of oxygen or for absorbing subsequent leakage during anoxia has been used up, there still must be 880 cc capacity of Ageless® always present so that leaking oxygen can be absorbed *fast* enough to be kept at a 0.1% oxygen level. (To maintain a level of 0.3%, only 290 cc of oxygen absorber would be needed.) The total calculated capacity needed to remove the oxygen from a 100 L enclosure (20.9 L of oxygen) plus that required to react with the 11 cc of oxygen per day leaking into the bag (11 × 14 days, or 154 cc) plus the amount of Ageless® needed to maintain the oxygen level at 0.1% just calculated (880 cc) is 21.9 L or 22 Ageless® Z-1000.

This result can be generalized: For a well-sealed enclosure with a moderately low leak rate, the "extra" amount of oxygen absorber needed to cope with the leak and also to maintain the oxygen level below 0.1% is less than a tenth of the absorber needed to react with the oxygen originally present in the bag. Fortunately, because of the excess over the nominal capacity that the manufacturers of oxygen absorbers place in each packet, the calculation of 22 Ageless® Z-1000 packets for this example is extremely conservative.

## Determination of the Oxygen Leak Rate of an Anoxic Enclosure

A leak test should first be performed on the anoxic enclosure before placing any objects in it in order to detect and correct any major leaks. For moderate to large bags or tents that are fitted with nitrogen exit and entry ports, preferably with valves on each as well as sensor connectors in the wall of the enclosure, go to Protocol B. (The nitrogen humidification step can be omitted.)

For those anoxic containers or bags without such nitrogen ports with valves, and thus requiring heat sealing of the unit during the test, start with Protocol A.

After completion of leak testing and corrective sealing, you should have at hand a portable oxygen analyzer that indicates a range of 0.1% to 1% oxygen with a sensitivity of about 0.02% and a remote oxygen sensor. (Placing the analyzer in the bag and omitting a remote sensor is possible if the batteries of the unit are adequate for continuous operation for at least two weeks.) A preferable alternative is the use of a trace oxygen analyzer such as the Teledyne T3000 with a remote sensor installed in the anoxic enclosure. Another alternative is a head-space analyzer like that described in Chapter 2 (p. 35) with a septum installed in the enclosure wall. This arrangement allows regular indication—and recording, with a datalogger—of the oxygen content inside. Daily piercing of the film to analyze the atmosphere within is equally acceptable if the piercing is made through a self-sealing septum.

The room in which the test is to be run should be at as constant a temperature as possible so as not to put any strain on the bag if the room becomes warmer and the bag expands. The next steps in determination of the leak rate are the same for any size of enclosure: (1) bring the internal atmosphere to an oxygen level less than about 0.3%; (2) close the valves to the inlet and outlet nitrogen lines, or seal the exit and the entrance holes; (3) wait half an hour or more until the oxygen values on the oxygen monitor stabilize; and (4) record the reading and time. Readings should be taken every day thereafter, making a neater record if taken at the same time each day. (The use of an instrument that has an RS 232 output is much preferred, of course. The data produced by it can be retrieved by a computer and a complete record of an entire period seen and manipulated in many forms.)

If the oxygen content of a 50 L bag changes by 50 ml of oxygen, that equals a change of 1000 ppm of oxygen and a change in the reading on the portable oxygen monitor of 0.1%. A 50 L bag made of Aclar (Filmpak 1177, FR7750: see Chapter 2, pp. 22, 23) would be expected to transmit about 6 to 10 cc of oxygen per day (about 200 ppm in a 50 L bag) through its area of film. Thus if the bag has been perfectly made and has no pinholes, five or six days may be required for a 0.1% increase to appear on the meter. If the maker of the bag has not had a fair amount of practice in heat sealing, half that time may be enough for a 0.1% change. In other words, unless the person who has fabricated it takes considerable care, coincidental and accidental leaks can equal the leak rate of the plastic laminate film from which the bag is constructed.

Occasionally, especially in the hands of inexperienced individuals, even an aluminized film like Marvelseal 360 (Chapter 2, pp. 21–23) in a 50 L or larger bag has a leak rate that results in a change of about 0.1% of oxygen in ten to fourteen days. (A pinhole can cause a leak of more than 1000 ppm per day.)

## Determination of the Rate of Reaction of Oxygen with Ageless®

A common question asked by conservators is, exactly how fast will an oxygen absorber remove oxygen from the air in the anoxic treatment? Before this question can be answered, one must know the size of the anoxic enclosure and how much absorber it contains. Given this information, it is still necessary to have the exact rate of reaction of oxygen with an absorber. That rate had to be determined at the GCI because it had not previously been determined.

The rate at which a chemical reaction proceeds is dependent on the concentration(s) of the reacting substance(s). In a simple form that is applicable to the oxygen–Ageless® reaction, the expression $[O_2]$ represents the concentration of oxygen in a sealed vessel and [Abs] the concentration of an oxygen absorber. Then the rate of reaction between the two to result in decrease in both would be

$$\text{Rate} = k\,[O2]\,[Abs] \qquad (1)$$

The concentration of oxygen can easily be found and expressed by the ppm of oxygen in the volume of the enclosure; that of the absorber is not so obvious. However, it was found possible to take it as the nominal oxygen-absorbing capacity of the packet(s) divided by the volume of the enclosure, [Abs] = C(capacity of absorber)/Vol. The value of k, $8.8 \times 10^{-3}$ per min, was determined by research at the GCI in more than one hundred experiments and a very wide range of chamber volumes (from 2 L to 1000 L) and ratios of oxygen absorber to volume (from 0.01 to 0.5) (Lambert, Daniel, and Preusser 1992). (Although most experiments involved consumption of only a portion of the oxygen absorber, many more recent results involving consumption of major fractions of the absorber have corroborated the value for k.)

Now the conservators' question can be answered. The rate of the reaction is the change in the oxygen concentration with time, that is, Rate = $d[O_2]/dt$. where t = time. If an example used a 100 L bag and oxygen absorber that could absorb 21 L (21,000 cc) of oxygen, how many cc of oxygen per minute would it absorb at the start? A simple plug-in to equation (1), Rate = k [O2] [Abs], gives the answer:

$$\begin{aligned} d[O2]/dt &= (8.8 \times 10^{-3})\,[O_2]\,[Abs] \\ [O_2]/\text{min} &= (8.8 \times 10^{-3}/\text{min})\,(209{,}000 \text{ ppm})\,(21 \text{ L}/100 \text{ L}) \\ &= (1839 \text{ ppm/min}) \times 0.21 \\ &= 387 \text{ ppm/min} \qquad (2) \end{aligned}$$

Because 1 ppm in a 100,000 cc bag is 0.1 cc, 387 ppm/min means that about 39 cc per minute of oxygen would react with this amount of oxygen absorber in this size of bag at the start of the anoxic treatment.

A usual follow-up question is, how long will it take a given amount of Ageless® or other oxygen absorber to change the normal 20.9% oxygen in the air of a particular bag down to 0.1% oxygen content? For simplicity, using the 100 L bag and 21 packets of Ageless® Z-1000 of the last example, the rate cannot continue as rapidly as it starts because the concentration of oxygen (and oxygen absorber) is constantly decreasing. Therefore, it would be improper just to divide the oxygen in the 100 L bag by 39 cc/min and come out with an answer of 2564 minutes or about 43 hours.

Instead the differential equation (2) is integrated to give

$$\ln \{[O_{2\ \text{starting}}] / [O_{2\ \text{at time t}}]\} = (8.8 \times 10^{-3})\ [\text{Abs}]\ t$$
$$\ln \{209{,}000\ \text{ppm}/1000\ \text{ppm}\} = (8.8 \times 10^{-3})\ (21/1000/\text{min})\ t$$
$$534 = (0.185/\text{min})\ t$$
$$t = 2886\ \text{min or } 48\ \text{hr}$$

The third kind of question that can be answered, how many oxygen absorber packets are needed to keep a chamber at 0.1% oxygen if one knows the leak rate of the chamber? has already been answered above. However, answering the same question for polyethylene shows why the use of such an oxygen-permeable film as polyethylene is inadvisable for anoxic enclosures: An enormous amount of oxygen absorber is required to maintain a 0.1% oxygen level.

That equation indicates that to ensure 0.1% (1000 ppm) oxygen in a 100 L polyethylene bag with polyethylene's leak rate of 800 cc (8000 ppm) per day would require the constant presence of 64,000 cc of nominal oxygen absorber capacity. In addition, because 800 cc of Ageless® capacity would be consumed by the transmitted oxygen each day, a further total of about 11,000 cc would be needed for fourteen days. This grand total requirement of 75,000 cc of Ageless® capacity—just for coping with the leakiness of polyethylene—indicates that using an inexpensive film does not result in an inexpensive anoxia process.

Considering only the oxygen leaking through the film and no accidental leaks, even a 100 L enclosure of Aclar with its relatively high leak rate (for oxygen-barrier films) of 11 cc per day requires 880 cc of oxygen absorber capacity to maintain a 0.1% oxygen level and consumes 154 cc of Ageless® capacity in fourteen days. This means that a total of only about 1030 cc of oxygen absorber is required to keep a low oxygen level and take care of leaks as compared to about 75,000 for polyethylene. Ageless® Z-1000 costs about $1 per packet from some dealers.

## References

**Lambert, R., V. Daniel, and F. Preusser**
1992. The rate of absorption of oxygen by Ageless®: The utility of an oxygen scavenger in sealed cases. *Studies in Conservation* 37:267–74.

# Appendix III

## Table of Conversion Factors

| To convert from | to | Multiply by |
|---|---|---|
| meter | inch | 0.3937 |
| cm | feet | 0.0328 |
| $cm^2$ | $inch^2$ | 0.155 |
| $cm^3$ | $inch^3$ | 0.0610 |
| $cm^3$ ( = cc = ml) | liter | 0.001 |
| feet | cm | 30.48 |
| feet | m | 0.3281 |
| $feet^2$ | $m^2$ | 0.0929 |
| $feet^3$ | $m^3$ | 0.0283 |
| $feet^3$ | liter | 28.3169 |
| gauge | mil | 0.01 |
| inch | cm | 2.54 |
| inch | m | 0.0254 |
| $inch^2$ | $cm^2$ | 6.4516 |
| $inch^3$ | liter | 0.0164 |
| $inch^3$ | $cm^3$ | 16.3871 |
| liter | $feet^3$ | 0.0353 |
| liter | $cm^3$ ( = cc = ml ) | 1000 |
| liter | $m^3$ | 0.001 |
| meter | inch | 39.3701 |
| meter | feet | 3.2808 |
| micron ($10^{-6}$ m) | mm | 0.001 |
| mm | micron | 1000 |
| micron | mil | 0.039 |
| mil | micron | 25.4 |
| $m^2$ | $feet^2$ | 10.7639 |
| $m^2$ | $inch^2$ | 1550 |
| $m^3$ | $feet^3$ | 35.3147 |
| $m^3$ | liter | 1000 |
| mil ( = 0.001 in.) | mm | 0.0254 |
| mil | gauge | 100 |
| mm | mil ( = 0.001 in.) | 39.3701 |
| percent | ppm | 10,000 |
| ppm | percent | 0.0001 |

*Note:* To convert from cc to ppm (e.g., for the oxygen in an anoxic enclosure : Divide cc by the volume of the enclosure (in cc), and multiply by one million.

To convert from ppm to cc: Divide ppm by one million, mutiply by volume of enclosure in cc.

# Appendix IV

## Manufacturers and Suppliers of Materials and Equipment Useful in Anoxia

The list of companies in this Appendix is not exhaustive. It is intended as a guide to sources that have been found helpful in our work in anoxia or have been recommended by others. Omission of any company or its products does not imply a negative judgment.

To avoid parochialism, we have tried to emphasize those companies that can serve the world community in two ways, either by direct sales via the Internet or through local suppliers. Therefore, we have provided the URL after the addresses and telephone numbers of all sources so that you can readily order items on-line or find the nearest dealers from a manufacturer's Home Page. The latter process is generally not easy in the case of large corporations such as Air Liquide or Air Products that are worldwide sources of gases. For their local distributors, it is easiest to look in your telephone book. It takes diligence, and perhaps an email to "Contact Us," to get information from the Web site of a megacorporation.

We have emphasized two sources, McMaster-Carr for a wide range of supplies and Cole-Parmer for scientific equipment, not to suggest that they are the only or even the best dealers on-line, but because searching their catalogs and ordering from them is especially easy. (Use of the "magnifying glass" icon on the toolbar to see the fine details of the object sought and of its printed specifications is surprisingly simple, even in the McMaster catalog, which is more than three thousand pages.)

Buying on-line is usually straightforward. Companies have told us that the best way to proceed as a new customer on the Web (assuming that explicit instructions are not given via logging on to the Web site) is simply to select the items you need as though you were an old customer. Then, when you wish to "check out," the screen may ask you to follow specific steps (sending or faxing a letter of credit or bank references, etc.) if they do not accept credit cards from new customers.

### Oxygen-Barrier Films, Bubbles, and Oxygen Absorbers

#### Oxygen-barrier films

The descriptions of oxygen-barrier films in Chapter 2 and the summary in Table 2.2 (p. 21) survey the most common films used in anoxia in museums. Our work has shown that even the Aclar type, with the "highest acceptable" oxygen transmission rate, can serve successfully in an anoxic treatment. Other people working with anoxia have reported that aluminized film with an extremely low OTR may have or may readily develop undesirable pinholes. Thus it appears that an exhaustive comparison of different film properties may be less necessary than care in fabrication and handling of the final enclosure. Local availability of a fairly good film may be preferable to a superb film that is either too expensive or too difficult to obtain.

The choice may be restricted by location or cost, but the outcome of the anoxic treatment, as in other procedures, can be mainly dependent on the user's attention to detail. The list below, condensed to a few representative films, follows the order of Table 2.2. The GCI value for the OTR is given, if available, as is the probable composition of the barrier layers in the laminate, the actual composition being proprietary.

ESCAL™, siloxane coated laminate, OTR 0.05, 0.11 mm (MGC, manufacturer)
  Conservation by Design, Keepsake, MGC, suppliers
Filmpak 1193, PVDC, OTR 0.28, 0.13 mm (Ludlow, manufacturer)
  Caltex, Edco., Techflex, suppliers
PTS, proprietary barrier film, OTR 0.5, 0.11 mm (MGC, manufacturer)
  MGC, Conservation by Design, suppliers
ArchiPress Pouches, PET, OTR 2.3
  Conservation by Design, supplier
BDF 2001, Cryovac type, EVA/EVOH (or EVAL), OTR 4, 0.03 mm (Cryovac, manufacturer)
  Metode, Conservation by Design, suppliers
Aclar type, CTFE, polyester, OTR 50, 0.11 mm (Honeywell, manufacturer)
  Aclar from Conservation Support, Filmpak 1177 from Caltex, FR 7750 from Techflex (suppliers)
Marvelseal 360, aluminized nylon, OTR 0.01, 0.014 mm (Ludlow, manufacturer)
  Caltex, Conservation Support, suppliers

## Bubbles, large flexible enclosures

Power Plastics, the manufacturer of CAT Bubbles, single-use aluminized PET film units, and multiple-use PVC enclosures, has a Canadian distributor, Maheu and Maheu, whose Web site has limited information and is solely in French. Thus it may be simpler to deal directly with the manufacturer.

## Oxygen absorbers and simple oxygen indicators

Ageless® and Ageless-Eye® (MGC, manufacturer)
  MGC, Conservation Support, Conservation by Design, University Products, suppliers
Atco™ (Standa, manufacturer)
  Metode, Standa, suppliers
FreshPax™ (Multisorb, manufacturer)
  IMPAK, Royco, suppliers

## Tools and Equipment

All the tools and equipment listed below have been mentioned in the previous chapters and most have been linked to a manufacturer or dealer. Here the numbers of sources have either been increased or dealers named where previously only the manufacturer was given. The order is approximately that of their presentation in the Protocols in Chapter 5.

### Leak detectors

Halogen leak detector—TIF Instruments, American Gas, local refrigerant supplies dealers
Ultrasonic leak detector (only for rigid anoxic chambers)—American Gas, supplier

### RH monitors

The most complete range of monitors is offered by a scientific apparatus dealer and manufacturer such as Cole-Parmer or Fisher Scientific or Omega. However, only a conservation supplier has simple relative humidity strips, as well as several types of hygrometers most useful for anoxia treatments.

Cole-Parmer, Fisher, Vaisala, Omega, Pave, Campbell, manufacturers
Conservation Support, Conservator's Emporium, Metode, University Products, suppliers

### Humidification systems

Fully automatic units that adjust the relative humidity of dry nitrogen are commercially available. They cost from $3,500 to $5,000.

Vacudyne, Power Plastics, Kojima Seisakujo, manufacturers
Metode, a Spanish supplier of anoxia materials produces a nonautomatic, less expensive humidification unit like that described in this book.

### Oxygen monitors

Wall-mounted safety monitors, with alarm; personal monitors with alarm—TIF, GasTech, GC Industries
Portable monitors, with integral or remote sensor (Teledyne)
Head-space analyzers, portable hand-held (PBI Dansensor)
bench type, with printer or RS 232 output (PBI Dansensor, Illinois)
Trace analyzers (Teledyne, Illinois, PBI Dansensor, Metode)

### Heat sealers

The simplest heat-sealing iron is essential even if the superior bar sealer is purchased because the latter is not always as effective for sealing inside corners or tight turns.

Heat-sealing iron, with thermostat and Teflon plate cover
McMaster-Carr, local photographic stores, Conservation by Design

Bar sealers, hand-operated and bench models
Conservation by Design (many types), Metode, IMPAK, McMaster-Carr

### Temperature sensors and monitors

Cole-Parmer, Vaisala, Omega Engineering, Conservation Support

### Dataloggers

Campbell, Mitchell, Pave, Logic Beach, Inc.
Conservation Support offers inexpensive units for anoxia use.

### Nitrogen sources and accessories

Gas in cylinders, liquid in dewars (Air Liquide, Air Products, Matheson, Praxair)
Nitrogen generators from compressed air (Air Products)
Two-stage regulators for cylinders and liquid
Local dealers of above nitrogen sources (Cole-Parmer)

### Miscellaneous

Many practical items, for example, small tools, wire, PVC pipe and fittings, paint, and sheets of plastic, can be purchased at local hardware stores or other nearby sources. Conservation suppliers, listed below, sell some of these items for anoxia. Beyond these, for an enormous range of both ordinary and hard-to-find hardware and workaday things, McMaster-Carr's 3000-page catalog is a rich source.

## Addresses and Web Sites

Air Liquide Group, www.airliquide.com. (The complexity of this Web site illustrates the difficulty of obtaining information quickly from some megacorporations. The small box at the lower left of the page referring to countries is useful for finding the Singapore Air Liquide nitrogen supplier immediately and useless for finding a dealer in the United States. The company, probably wisely, expects one to use the local telephone book.)

Air Products and Chemicals, Inc., 7201 Hamilton Blvd., Allentown, PA 18195; (610) 481-4911; Fax (610) 481-5900; www.airproducts.com. (The best source of safety information about gaseous and liquid nitrogen.)

American Gas and Chemical Company, Ltd., 220 Pegasus Ave., Northvale, NJ 07647; (800) 288-3647; Fax (201) 767-1741; www.amgas.com.

Bemco Inc., Environmental & Space Simulation Systems, 2255 Union Pl., Simi Valley, CA 93065; (805) 583-4970; Fax (805) 583-5033; www.bemcoinc.com.

CalTex Plastics, 2110 East 51st St., Vernon, CA 90058; (800) 584-7303; Fax (323) 583-1207; www.caltexplastics.com.

Campbell Scientific Inc., 815 West 1800 N., Logan, UT 84321; (435) 753-2342; Fax (435) 750-9540; www.campbellsci.com.

Cole-Parmer Instrument Company, 625 E. Bunker Ct., Vernon Hills, IL 60714; (800) 323-4340; Fax (847) 247-2929; www.coleparmer.com.

Conservation by Design Ltd., Timecare Works, 5 Singer Way, Woburn Road Industrial Estate, Kempston, Bedford MK42 7AW, UK; (01234) 853-555; Fax (01234) 852-334; www.conservation-by-design.co.uk.

Conservation Support Systems, 924 W. Pedregosa St., Santa Barbara, CA 93101; (800) 482-6299; Fax (800) 605-7503.

Conservator's Emporium, 100 Standing Rock Circle, Reno, NV 89511; (775) 852-0404; Fax (775) 852-3737; www.consemp.com.

Cryovac Division, Sealed Air Corporation, Park 80, East Saddlebrook, NJ 07663; (201) 791-7600; www.sealedair.com.

Edco Supply, 323 36th St., Brooklyn, NY 11232; (718) 788-8108, Fax (718) 788-7481; www.edcosupply.com.

Fisher Scientific, 2000 Park Lane, Pittsburgh, PA 15275; (800) 766-7000; Fax (800) 926-1166; www.fishersci.com. (Large supplier of scientific equipment.)

Gas Tech, Inc., 8407 Central Ave., Newark, CA 94560; (510) 745-8700; Fax (510) 794-6201; www.gastech.com.

GC Industries, 2991 Corvin Dr., Santa Clara, CA 95051; (408) 737-0550; Fax (408) 737-0346; www.gcind.com.

Honeywell Industrial Automation and Control, 16404 N. Brack Canyon Hwy, Phoenix, AZ 85023; (602) 863-5000; Fax (800) 343-0228; www.industrialstore.honeywell.com.

Illinois Instruments, Inc., 27840 Concrete Dr., Ingleside, IL 60041; (815) 344-6212; Fax (815) 344-6332; www.illinoisinstruments.com.

IMPAK Corporation, 419 Del Monte, Pasadena, CA 91103; (626) 398-7300; www.sorbentsystems.com.

Keepsafe Systems, Inc., 570 King St. W.,Toronto, Ont. M5V 1M3, Canada; (800) 683-4696, Fax (416) 703-5991; www.keepsafe.ca.

Kojima Seisakujo, Inc., 6F, Popura Building, Ohdenma-cho, Nihonbashi, Chuo-ku,

Tokyo 103; 81-3-3664-02000; Fax 81-3-3664-0210; www.kofloc.co.jp.

Ludlow Laminating & Coating Division, Tyco Industries, 4058 Hwy 79, Homer, LA 71040; (318) 927-2531; Fax (318) 927-9825; www.tycoplastics.com.

Logic Beach, Inc., 8363-6F Center Dr., La Mesa, CA 91942; (619) 698-3300; Fax (619) 469-8604; www.logicbeach.com.

Maheu & Maheu, Inc., 710 Bouvier, Bureau 195, Québec G2J 1C2 Canada; (418) 623-8000; (800) 463-2186; Fax (418) 623 5584; www.maheu-maheu.com.

Matheson Tri-Gas, Inc., P.O. Box 624, 959 Route 46 East, Parsippany, NJ; (800) 416-2505; www.mathesontrigas.com.

McMaster-Carr Supply Company, 9630 Norwalk Blvd., Santa Fe Springs, CA 90670; (562) 692-5911; Fax (562) 695-2323; www.mcmaster.com.

Metode, Conservation and Exhibition Systems, Ronda del Guinardo, 59 bis, 08024 Barcelona, Spain; 34-3-446-37-10; Fax 34-3-446-37-15; metode@deinfo.es www.metode.net.

Mitchell Instrument Co., 1570 Cherokee St., San Marcos, CA 92069-6901; (760) 744-2690; Fax (760) 744-0083.

Mitsubishi Gas Chemical America, Inc., 520 Madison Ave., 17th Fl., New York, NY 10022; (212) 752-4620; Fax (888) 909-6422; www.mgc-a.com.

Multisorb Technologies, Inc., 325 Harlem St., #24, Buffalo, NY 14224; (716) 824-8900; Fax (716) 824-4091; www.multisorb.com.

Omega Engineering PCP, Inc., One Omega Dr., Stamford, CT 06907-0047; (203) 359-1660/(800) 848-4286; Fax (203) 359-7700; www.omega.com.

Pace Scientific, 6407 Idlewild Rd., Suite 2214, Charlotte, NC 28212; (704) 568-3691; Fax (704) 568-0278; www.pace-sci.com.

Pave Technology Co., Inc., 2751 Thunderhawk Ct., Dayton, OH 45414; (937) 890-1100; Fax (937) 890-1592; www.pavetechnologyco.com.

PBI-Dansensor America, Inc., 139 Harristown Rd., Glen Rock, NJ 07452; (201) 251-6490; Fax (201) 251-6491; www.pbi-dansensor.com.

Power Plastics, Station Road, Thirsk, North Yorkshire YO7 1PZ, UK; (01845) 25503; Fax (01845) 525 485; www.powerplastics.co.uk.

Praxair Technology, Inc., www.praxair.com. (International supplier of gases and liquefied gases.)

Preservation Equipment, Ltd., Vinces Road, Diss, Norfolk IP22 4HQ, UK; (01379) 647 400; Fax (01379) 650 482; www.preservation equipment.com. (Preservation Equipment is represented in North America by University Products, www.universityproducts.com.)

Royco Packaging, 3979 Mann Rd., Huntingdon Vally, PA 19006; (877) 645-7732; Fax (215) 322-9260; www.roycopackaging.com.

Sealed Air Co., Park 80 East, Saddle Brook, NJ 07663; (201) 791-7600; Fax (201) 703-4205; www.sealedair.com.

Shannon Packing Co., 3647 Shasta St., Chino, CA 91710; (909) 591-8768; Fax (909) 591-1409.

Standa Industrie, 68 Rue Robert Kaskoreff, 14050 Caen, France; 31745489; Fax +33 2 3173 1357; www.standa-fr.com; www.atmosphere-controle.fr/.

Swagelok Co., 6090 Cochran Rd., Solon, OH 44139; www.swagelok.com/distributors.

Techflex Packaging, 13826 S. Prairie Ave. Hawthorne, CA 90250; (310) 973-1321; Fax (310) 973-2636; www.xsential.com.

Teledyne Analytical Instruments, 16830 Chestnut St., City of Industry, CA 91749; (626) 961-9221; Fax (626) 961-2538; www.teledyne-ai.com.

Thermotron Industries, 291 Kollen Park Dr., Holland, MI 49423; (616) 393-4580; Fax (616) 392-5643; www.thermotron.com.

TIF Instruments, Inc., 9101 NW 7th Ave., Miami, FL 33150; (800) 327-5060/(305) 757-8811; Fax (305) 757 3105; www.tif.com.

University Products, Inc., P.O. Box 101, 517 Main St., Holyoke, MA 01041-0101; (800) 628-1912; Fax (800) 532-9281; www.universityproducts.com.

Vacudyne, Inc., 375 Joe Orr Rd., Chicago Heights, IL 60411; (708) 757-5200, ext. 226; Fax (708) 757-7180; www.vacudyne.com.

Vaisala, Inc.; Sensors Systems Div., 100 Commerce Way, Woburn, MA 01801-1068; (781) 933-4500; Fax (781) 933-8029; www.vaisala.com.

# Appendix V

## Outlines for Treatment Protocols A, B, and C

### Outline for Protocol A.
### Anoxia in Small to Medium-Size Bags or Pouches

1. Obtain materials and equipment:
    - film
    - knife and scissor
    - heat sealer
    - oxygen absorber
    - RH strip or hygrometer
    - silica gel
    - leak detector and tracer gas
2. Provide protection if objects are fragile.
3. Measure and cut the plastic film to the right size for the bag.

[4. Only if you chose aluminized film for the bag, seal a window in the bag.]

5. Form two sides of the bag by sealing. Insert units for anoxia treatment.
6. Finish sealing the bag. Test for leaks.
7. Check the relative humidity, temperature, and oxygen level inside the bag for fourteen days.
8. Open the treatment bag. Remove the oxygen absorber and properly dispose of it.

### Outline for Protocol B.
### Anoxia in Large User-Made Tents

1. Obtain materials and equipment:
    - film
    - knife, scissor, and punch
    - a set of wrenches
    - heat sealer
    - grease
    - pipes for frame and lifting device
    - saw or cutter for cutting pipes
    - rope and pulley for the lifting device
    - copper or nylon tubes and Swagelok or airtight tube fittings
    - humidification unit

- large nitrogen supply with a regulator
- vacuum pump
- flowmeters
- oxygen, relative humidity, and temperature sensors and displays or a datalogger
- leak detector and tracer gas

2. Prepare the base.
3. Construct a lifting device for the tent and an internal frame for its support.
4. Fabricate the tent.
5. Install tent, lifting device, and connectors. Place self-adhesive strips on floor.
6. Complete tent installation. Calibrate monitor, start leak test, attach humidifier.
7. Pressurize tent slightly with humidified nitrogen. Finish leak test. Seal any leaks.
8. Purge the tent of oxygen.
9. Begin treatment. Repurge when or if necessary.
10. Complete anoxia process. Release nitrogen to outdoors.

### Outline for Protocol C.
### Anoxia in Large Commercial Enclosures, "Bubbles"

1. Obtain materials and equipment:
   - commercial Bubble
   - knife, scissor, and punch
   - small drill and drill bit
   - a set of wrenches
   - grease
   - pipes for frame
   - saw or cutter for cutting pipes
   - copper or nylon tubes and swagelok or air-tight tube fittings
   - humidification unit
   - large nitrogen supply with a regulator
   - vacuum pump
   - flowmeters
   - oxygen, relative humidity, and temperature sensors and displays or a datalogger
   - leak detector and tracer gas
2. Prepare the base.
3. Make an internal frame for the Bubble. Install nitrogen, electrical connectors.
4. Put infested objects in the Bubble area. Set the Bubble top on the frame.
5. Attach the humidified nitrogen system, test oxygen monitor, start leak test.
6. Pressurize Bubble slightly with nitrogen. Perform leak test. Seal leaks.

8. Purge the tent of oxygen with humidified nitrogen.
9. Begin treatment. Repurge when or if necessary.
10. Complete anoxia process. Release nitrogen to outdoors.

# Index

Note: Page numbers followed by the letters *f* and *t* indicate figures and tables, respectively.

## About the Authors

**Shin Maekawa** received his M.S. degree in mechanical engineering from the University of California, Los Angeles, and has been a professional engineer for the State of California. He is currently a senior scientist at the Getty Conservation Institute. His research focuses on oxygen-free environments for the control of insect pests and long-term storage and display of artifacts, design of specialized storage and display cases, and monitoring and control of environments for artifacts as well as historical buildings.

**Kerstin Elert** received her diploma in objects conservation from the Staatliche Akademie der bildenden Kunste in Stuttgart, Germany. She was a research fellow at the Getty Conservation Institute, where she conducted research on anoxic treatment for insect control between 1995 and 1998. Currently she is a research fellow at the Department of Mineralogy and Petrology at the University of Granada.